useless arithmetic

Why Environmental Scientists

Columbia University Press   New York

useless arithmetic

Can't Predict the Future

Orrin H. Pilkey & Linda Pilkey-Jarvis

Columbia University Press
Publishers Since 1893
New York   Chichester, West Sussex
Copyright © 2007 Columbia University Press
All rights reserved

Library of Congress Cataloging-in-Publication Data
Pilkey, Orrin H., 1934–
Useless arithmetic : why environmental scientists can't predict the future / Orrin H.
Pilkey and Linda Pilkey-Jarvis.
cm.
Includes bibliographical references and index.
ISBN 978-0-231-13212-1 (cloth)
1. Environmental protection—Decision making—Mathematics. 2. Environmental
policy—Mathematics. 3. Ecology—Mathematical models. I. Pilkey-Jarvis, Linda. II. Title.
TD171.8.P55   2006
363.7'05—dc22
2006009632

∞

Columbia University Press books are printed on
permanent and durable acid-free paper.

This book was printed on paper with recycled content.

Printed in the United States of America
c 10 9 8 7 6 5 4 3 2

To Jim, Sharlene, and Eddie

contents

preface

According to Greek mythology, Zeus once released two eagles in order to find the center of the earth. One flew east and the other west. The birds met at Delphi, which lies on the slopes of Mount Parnassus. From about 1400 B.C. to A.D. 381, the Oracle of Delphi held sway at what was the most important shrine in all of Greece. The oracle could be more accurately described as a succession of priestesses, each given the title of Pythia. For twelve centuries the oracle played an influential role in ancient history and determined the course of empires.

Built around a sacred spring, the shrine to the oracle attracted people from all over Greece and far beyond, who came to pose their questions about the future to the Pythia. Her cryptic answers covered everything from optimal sowing and harvesting times to when an empire should declare war. As she responded to questions, seemingly in a trance, her inarticulate cries were interpreted and written down by an official scribe. In early times this transcription was rendered in hexameter verse, but later it was written in prose. The priest Plutarch said that the trance was the result of vapors, and indeed this may have been the case, for according to a recent geologic study, the presence of ethylene gas (once used as an anesthetic) has been detected in the vicinity of the spring.

The oracular responses were notoriously ambiguous, and their interpretation was often "deduced" only after the event to which they referred. Arguments over the correct interpretation of an oracle were common, but the oracle could always clarify or give another prophecy if more gold was provided. A good example is the incident before the Battle of Salamis, in which the Greeks defeated the Persians. The Pythia first predicted doom and later predicted that a "wooden wall" (interpreted by the Athenians to mean their ships) would save them.

Fast-forward 2,300 years and we find a world that still highly values and relies on prediction. Modern-day oracles are expected to provide predictions over a much wider range of things than the Oracle of Delphi could ever have imagined. In fact, with all the politicians, pundits, government agencies, stockbrokers, scientists, and academics offering their views today, we citizens are inundated with advice and suggestions derived from predictions about the future.

One type of prediction that the original Pythia seldom had to worry about has to do with processes on the surface of the earth. During the time of the Pythia, the earth was far less densely populated, and society had fewer machines to move soil, fight wars, or pollute the air and water. In the days of the American frontier you could start excavating a mine shaft in Montana whenever you wished, provided you could file the claim and pay for the dynamite. If you could make or buy a boat, all the fish in the sea were yours, provided you could catch them. And if you had an eroding shoreline in front of your house, you could build a seawall at will or dump a few dozen truckloads of sand or construction debris on the beach.

Times have changed. Before we can develop a new mine now, a vast amount of paperwork is required, including an environmental impact statement. Such statements are predictions of the ways in which the proposed project could affect the quality of air and water in the neighborhood, and the quality of life for plants and animals and humans alike. Shored up by the cries of distress from the mostly wealthy people who live next to beaches, the federal government began funding beach nourishment projects on Great Lakes and ocean shorelines. In order for a community to receive federal funding for an artificial beach, the calculation of a cost-benefit ratio is required, which in turn assumes an accurate prediction of how rapidly the artificial beach will disappear. Shock waves from the demise of the Grand Banks cod fishery, perhaps the world's greatest fishery for more than five hundred years, have bolstered the requirements for accurate estimates of fish stocks as a basis upon which to regulate fishing.

The widespread availability of computers, the requirement for environmental impact statements and cost-benefit ratios, and the dawn of mathematical models all arrived on the scene simultaneously in the final quarter of the twentieth century. Scientists in the 1960s and 1970s assured bureaucrats that the computer would make it possible to predict the outcomes of natural processes accurately. We don't know how to do it right now, they said, but fund us and we'll figure it out. There are still some scientists who claim successes—undaunted by several decades of the failure of certain mathematical models to provide the accurate answers that society needs.

At the beginning of the twenty-first century, predictive models of processes on the surface of the earth have come into widespread use. The recognition of complexity and chaos seems not to have diminished the still-rising star of modeling. Every year hundreds of cost-benefit ratios roll off the presses for federal engineering projects involving beaches, rivers, lakes, and groundwater flow. Engineers who have found great success in the use of models to predict the behavior of steel and concrete have applied modeling to the natural environment just as if nature were made up of construction materials with well-defined properties.

The environmental impact of various engineering activities 50 years into the future is calculated even more frequently than cost-benefit ratios are. The mother of all environmental impact predictions is the required assurance of 10,000 years of safety from the Yucca Mountain repository of the nation's radioactive waste. Billions of dollars have been spent at Yucca Mountain on the unrealistic goal of predicting what the climate and groundwater flow will be thousands of years from now. The American judiciary apparently is even more clueless than the scientists of the Department of Energy who are charged with proving the safety of Yucca Mountain—recently a federal court decreed that the prediction must cover 300,000 to 1 million years! The *New York Times* quotes an incredulous bartender in Las Vegas as saying, "The earth might not even be here a million years from now." The disappearance of the earth is perhaps not likely, but certainly over the next several hundred thousand years there will be two or three ice ages, the sea level will fall and rise by hundreds of feet, and Yucca Mountain will experience major changes in climate, perhaps an earthquake or two, maybe even a volcanic eruption. Undying faith in mathematics stilled the voice of scientific caution and skepticism that should have warned Congress and the judiciary that the predictive requirements they established for a repository at Yucca Mountain were impossible to achieve.

The reliance on mathematical models has done tangible damage to our society in many ways. Bureaucrats who don't understand the limitations of modeled predictions often use them. That was why the Bureau of Land Management allowed open-pit mines that, once abandoned, would eventually become "giant cups of poison." Models act as convenient fig leaves for politicians, allowing them to put off needed action on controversial issues. Fishery models provided the fig leaf for Canadian politicians to ignore the dying Grand Banks cod fishery. Agencies that depend upon project approvals for their very survival (such as the U.S. Army Corps of Engineers) can and frequently do find ways to adjust models to come up with correct answers that will ensure project funding. Most damaging of all is the unquestioning acceptance of the models by the public because they are assured that the modeled predictions are the state-of-the-art way to go.

If all this is true, how can people counteract the modeling craze? The supposition is that there is no way that ordinary people can argue with such sophisticated mathematics. But there is more to models than mathematics. There are parameters such as water velocity, temperature, wave height, rock composition and porosity, and many other factors that make natural processes work. And each of the parameters is represented in a model by simplifications and assumptions. This is the point at which the mathematically challenged among us can evaluate models and even question the modelers.

For example, the height of the waves striking a beach is an important control on the velocity of currents that carry sand away. Anyone who has spent time on a beach, however, knows that the waves vary widely from day to day and, of course, during a storm can be huge. So what number do you use in a model to represent such a variable parameter? The volume and flow rate of groundwater is an important factor in controlling the fate of nuclear waste at Yucca Mountain, Nevada, and the amount of rainfall will be critical in determining that rate. What number do you use in the model for the annual rainfall 100, 1,000, 10,000, or 1,000,000 years from now? After an open-pit mine is abandoned, the rate of flow of groundwater into the pit is critical to understanding whether or not the pit will be an environmental hazard, but the rate of flow into the pit will vary as acidic waters either dissolve rock and enlarge pores or precipitate minerals and reduce pores. Future rainfall amounts are also important.

How do you put all of this together and come up with a prediction of the composition of the pit lake 50 years from now? Or 100 years from now?

Years ago, in his capacity as a professor at Duke University, Orrin organized a graduate seminar in the Nicholas School of the Environment to look at mathematical models used in coastal geology. None of the class participants (including the professor) knew much about mathematical models. They decided to get to the bottom of the question of why the models seemed to come up with inaccurate predictions of the behavior of beaches.

What a revelation that seminar turned out to be! It became clear that beach modelers used models that had no demonstrable basis in nature. They employed "coefficients" that in reality were fudge factors to assure that the "correct" answer would be found, and no one looked back to see if the models actually worked. And no one was complaining. Neither the public nor the politicians knew or particularly cared, since the models were providing them with federal funds to stop beach erosion. And when the scope of the seminar was broadened beyond beaches, it became apparent that the problem existed in a wide variety of modeling efforts involved with all kinds of physical and biological processes concerned with the surface of the earth.

Clearly, the mathematical modeling community believed so strongly in models that it insisted on using them even when there was no scientific basis for their application. The discredited Bruun Rule model predicts how much shoreline erosion will be created by sea-level rise, and since no other model claims to do this, the Brunn Rule remains in widespread use. The maximum sustainable yield is a concept that fishery models are still using as a means to preserve fish populations despite the fact that the concept was discredited thirty-five years ago.

Participants in the seminar came to believe that an amazing statement by Jim O'Malley, a representative of the fishing industry, could be applied on a much broader front than fish models:

> I stress that the problem was not mathematics per se but the place of idolatry we have given it. And it is idolatry. Like any priesthood, it has developed its own language, rituals and mystical signs to maintain its status and to keep a befuddled congregation subservient, convinced that criticism is blasphemy. . . . Most frightening of all, our complacent acceptance of this approach shows that mathematics has become a substitute for science. It has become a defense against an appropriate humility, and a barrier to the acquisition of knowledge and understanding of our ocean environments. . . . When used improperly, mathematics becomes a reason to accept absurdity.

Linda has worked for both federal and state governments. Quantitative modelers, she independently observed, have an almost religiously fanatic outlook on the veracity of their models and brook little criticism. It is a characteristic we believe can be applied broadly to many natural-process modelers. The modeling modus operandi is shrouded in mystery, with necessary though poorly communicated assumptions made at each step along the way. In Linda's view, those who rely on the models for making policy decisions rarely understand the limitations of the models, much less are prepared to communicate such information to the public.

Qualitative models are used in trying to understand natural processes; here precise answers are not sought. Such models seek only trends, relative impacts, probable causes, directions of flow, timing of events. They consider and incorporate only the most important parameters of a process. They are not expected to produce accurate answers. These models often work and can be very useful. In this book we are concerned with the quantitative, "accurate" predictions made by mathematical models that are applied to societally important issues involving natural surface events on the earth. These models are expected to produce answers that are accurate enough to use for engineering and other applied societal purposes.

The book is intended to be read by non-specialists who are interested in nature and in the politics of working with the earth. We have not included equations here except (with some reluctance) for a few relatively simple examples in an appendix. Without resorting to mathematics, we make our point that applied quantitative mathematical models of earth processes cannot produce accurate answers. We evaluate assumptions behind the models, look at the nature of the field data that go into the models, evaluate model achievements, and examine the dialogue between modelers and their "customers." We are speaking to non-mathematicians like ourselves.

In the process of writing this book we received many ideas and much encouragement from the small but growing group of those who are skeptical about earth surface process modeling. Probably more than any other individual, Peter Haff, a colleague of Orrin's on the Duke University faculty, provided the impetus, encouragement, and education that we needed to move ahead with this book. He won't, however, agree with everything we have said here! Art Trembanis, a professor at the University of Delaware, provided valuable insights into the philosophy of science and modeling.

Hours and hours of discussions about models with our friend Andrew Cooper of the University of Ulster in Northern Ireland produced a lot of additions and revisions for our project. Andrew was the one who alerted us to the crisis of modeling ensconced within the crisis of AIDS in Africa. Rob Young, Rob Thieler, and David Bush, all special friends, provided endless discussions that brought life to a number of the book's chapters. Over the last five years or so, we have, at the drop of a hat, discussed mathematical modeling with a large number of people. These include Paul Baker, Victor Baker, Ron Brunner, Brad Murray, Michael Orbach, Roger Pielke, Walter Pilkey, Cathy Rigsbee, Daniel Sarewitz, and Jordan Slott. Columbia University Press editors Robin Smith and Patrick Fitzgerald were most helpful with comments and suggestions along the path of writing. We are particularly grateful to the anonymous reviewer who seems to have read every word in the book and made numerous thoughtful recommendations. Copy editor Jan McInroy edited out a lot of fuzzy wording and strange punctuation. Andy Coburn, who occupies the office next to Orrin's, constantly advised him on the vagaries of computers and the mysteries of Googling. Chapter by chapter, a number of people read our individual sections or portions thereof. We hasten to note that not all agreed with each of our points; mathematical modeling criticisms bring out a wide variety of viewpoints and emotions. Following is a list of those who read individual chapters or who offered substantial advice that guided our thinking.

Chapter 1—James Wilson, Kathy Dixon, David Rackley, Peter Haff, Jim O'Malley, Corey Dean, Robin Smith; chapter 2—Rob Young, Peter Haff, William Neal, Art Trembanis, Jordan Slott, Wallace Kaufman, Walter Pilkey, Diane Pilkey, Keith Pilkey; chapter 3—Norm Christiansen; chapter 4—Kathy Dixon, Ron Brunner, Paul Baker, Gabriele Hegerl, Art Trembanis, Robin Smith; chapter 5—Andrew Cooper, Diane Pilkey, Kathy Dixon, Keith Pilkey, Robin Smith; chapter 6—Rob Young, Art Trembanis, Robin Smith; chapter 7—Bob Moran, Tom Myers, Glen Miller, Wally Kaufman, Kathy Dixon; chapter 8—Sylvan Kaufman; chapter 9—Art Trembanis, Keith Pilkey, Diane Pilkey.

Welcome to the world of mathematical models. We hope that after reading this book you will view these ever more important tools of science through different eyes.

Whenever I hear a fishery scientist proclaim that his analysis is rigorous, I am reminded about what John Kenneth Galbraith is reputed to have said once to a group of economists: that the prestige of mathematics has given economics rigor but, alas, also mortis.

—Jim O'Malley, fishing industry representative and executive director of the
East Coast Fisheries Federation

mathematical fishing

The Almighty Cod

More than five hundred years ago, fishers from Portugal and the Basque region of Spain began fishing the fabled Grand Banks of Canada. Although many species of fish were harvested from the seemingly inexhaustible stock, the most famous and valuable was the cod. Thousands of vessels sailed back to Spain and Portugal, from the New World to the Old, their holds jammed with barrels of salted cod. Codfish—*bacalao* in Spain and *bacalhau* in Portugal—became a food staple for the entire Iberian Peninsula. Salted cod achieved added importance because of the numerous meatless days imposed by the Catholic Church. Later, generations of North American children learned of the importance of another cod product, the foul-tasting cod liver oil valued (by parents) as a source of vitamin D.

The Grand Banks are on the Canadian continental shelf off Newfoundland (figure 1.1). Nearly 300 miles across, it is one of the widest continental shelves of the world. The banks cover an area of 110,000 square miles and consist of shallow submarine plateaus, 75 to 300 feet

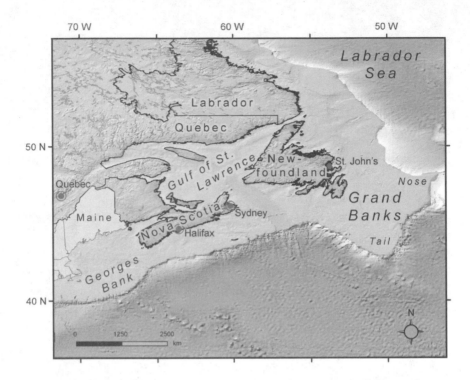

Figure 1.1 This physiographic diagram of a portion of the North American continental margin shows the Grand Banks and Georges Bank, both very important fishing grounds. In 1992 the cod fishery on the Grand Banks crashed as a result of overfishing, and it has not recovered since. Mathematical models must bear some of the blame for this failure of what may have been the world's richest fishery. Cod are still harvested from Georges Bank, but in much smaller numbers that in previous years. Map by David Lewis.

deep, separated by troughs that are 600 or more feet deep. The cold Labrador Current flows down from the north, to mix over the banks with the warm Gulf Stream coming up from the south. The resulting churned-up waters are rich in nutrients and support a huge marine ecosystem. Icebergs are commonly present, slowly drifting south, melting along the way. The winter storms on the banks are legendary, but the water never freezes over.

The Atlantic cod, *Gadus morhua* (figure 1.2), has always been the mainstay of the Grand Banks fishery. Perhaps 90 percent of the fish catch on the banks during the 1980s was cod. It is a tasty fish that can be salted or sun-dried and preserved for a long time, which was of particular importance in the days before refrigeration. Cod is often the fish used for fish-and-chips and for the McDonald's fish sandwich.

Cod have an olive green spotted back and a white belly, with a prominent, slightly curved back-to-front stripe along the side. Various shades of brown and even red may be present, depending upon the habitat. They are commonly two to three feet long and weigh five to ten pounds, although occasionally in the past individual fish "as big as a man," six feet long and two hundred pounds, were caught. They continue to grow during their entire lifetime.

Cod were once found in schools, sometimes miles across, in deep water in the winter and in shallower water in the summer. The Atlantic cod probably has a number of subpopulations, each following the same migration paths year after year. The Northern cod used to extend from off the tip of Labrador down to Cape Hatteras.

The cod eats just about anything, including the occasional unwary seabird resting on the rolling ocean surface. It is a fish that virtually swims with its mouth open, devouring clams, squid, mussels, echinoderms, jellyfish, sea squirts, worms, and other fish, including its own young. Its favorite fish is perhaps the capelin, a small plankton feeder that spawns in the summer on and near beaches. Capelin are probably responsible for the cod's migration to shallow water in the summertime. Many who have written about the demise of the Atlantic cod have noted the irony that a fish as greedy as the cod is being destroyed by humans, another of God's creatures with even greater greed.

Cod spawn between March and June, releasing eggs that float to the surface and become part of the plankton for ten weeks. When the larvae reach one inch in length, they swim back to the bottom. Each female cod releases between 2 million and 11 million eggs—a stupendous figure

Figure 1.2 The Northern cod (*Gadus morhua*), shown here, was once the mainstay of the world's greatest fishing grounds, the Grand Banks of Canada. Misplaced confidence in mathematical models played a role in the demise of this fishery. Drawing courtesy of the National Oceanic and Atmospheric Administration; modified by Dave Lewis.

that gave rise to the poem (said to be written by an anonymous American) comparing the productivity of codfish and chickens:

> The Codfish lays 10,000 eggs
> The lowly hen but one;
> But the codfish never cackles
> To tell what she has done.
> And so we scorn the codfish
> While the humble hen we prize,
> Which only goes to show you
> That it pays to advertise.

For hundreds of years, Grand Banks fishers caught cod from small dories manned by one or two men, using herring-baited hooks. The boats were lowered from a mother ship each morning and gathered back in by nightfall. It was dangerous work immortalized by Winslow Homer's famous painting *Lost on the Grand Banks*. The seascape shows two forlorn fishers, separated from the mother ship, peering over the side of their dory in rough weather. Microsoft mogul Bill Gates purchased the painting in 1998 for $30 million. It was, by a factor of three, the highest price ever paid for an American painting.

Gradually, newer and more efficient fishing methods came along (figure 1.3), especially in recent decades. These include nearshore traps, used when cod come in to shallow water during the summer. Seines, or nets pulled into circular traps by small motor vessels, and untended drift nets are also both used on the Grand Banks. In some fisheries (not cod), drift nets can be as long as forty miles.

This method of cod fishing has been a particularly insidious and wasteful killer of Grand Banks fish. When the nets are lost or untended, large numbers of fish are caught by their gills as the net eventually sinks to the seafloor. Scavengers empty the net, which once again floats to the surface, fills with fish, sinks, and again returns to the surface after being emptied. The deadly cycle continues until the net disintegrates, which may take years if it is made of durable nylon.

But the biggest problem for the cod fishery on the Grand Banks was the fishing trawlers. These vessels drag nets shaped like giant bags behind them, scooping up everything in their path. Invention of the *otter trawl*, which uses chain weights to hold the net on the bottom and "doors" attached to the towing cables that keep the net open, was a major step in the evolution of trawls. The otter trawl makes it possible to drag nets over

Figure 1.3 Hand-line fishing for cod in rough weather on the Georges Bank, from a painting by Paul E. Collins. It was a rugged life for those who went to sea in these cold, rough waters! Courtesy of the National Oceanic and Atmospheric Administration.

uneven bottoms. Later, the invention of electronic devices that could spot fish schools and guide the towing vessel in their direction added more efficiency to the fishery. And then other devices told the trawler skipper when the bag was full, preventing a premature retrieval or loss of the catch if an overfull bag broke while being hauled on board.

In the mid-1980s, rock hopper dredges came in. These are trawl nets with large, heavy wheels capable of rolling over almost any seafloor obstructions and preventing the net from being torn. Bottom creatures of all kinds, often with no food value, are captured or scraped away. These modern trawlers, if not regulated in some way, can take more fish than the fishery can sustain.

Overfishing or not, they can destroy the very environment needed for recruitment of the next generation of fish. Studies have shown that juvenile cod survive best in areas that have rough bottoms, hiding from predators behind and within the many nooks and crannies afforded by such a seafloor. Almost any seafloor irregularity can provide shelter—rocks, shells, ripple marks, mud patches, sponges, worm tubes, depressions excavated by fish and rays. Since an area equal to all of the world's continental shelves is trawled every two years, the habitats provided by an uneven seafloor disappear into geologic history.

The Grand Finale of the Grand Banks

With the benefit of impeccable hindsight, it is possible to watch with equal parts fascination and horror as an entire industry and ecosystem drives off a cliff. In 1968 the cod catch on the banks was 810,000 tons. The total cod catch from the Grand Banks, the Bay of Fundy, and the Gulf of St. Lawrence reached 1,900,000 tons! Pol Chantraine, in his book *The Last Codfish*, called it "senseless, wild over fishing." In 1992 the Grand Banks fishery collapsed; it was the biggest fishery disaster ever. The cod and flounder were no more; they had joined the haddock fishery that had already crashed in the 1950s and never recovered. Forty thousand jobs and a way of life disappeared as the world's most famous fishing grounds closed up. People, young and old, began to leave the remote fishing villages lining the shores of Newfoundland, hoping for greener pastures on the mainland. More than a decade later, the cod have still not recovered, perhaps because of the effect known as *dispensatio*, a lowering of reproduction rates that occurs when the density of a fish population is no longer sustainable. Among other things, dispensatio reflects the difficulty that widely dispersed fish have in finding one another for mating. So few cod are left that now there is a push to have the Northern cod declared an endangered species!

How could the Grand Banks, with all of its high visibility, go belly-up? Or, as writer Deborah MacKenzie in the 1995 *New Scientist* asks: "How could an advanced nation with an army of scientists allow one of the richest fisheries in the world to . . . be destroyed?" The *Boston Globe* noted that "after five centuries of abundance, the cod are gone from the Grand Banks of the North Atlantic, wiped out by a combination of scientific mismanagement, bureaucratic sloth and above all, almost incomprehensibly mindless greed." Richard Cashin, chairman of a task force looking into the collapse, characterized it as "a famine of biblical scale—a great destruction."

Overfishing, poor fishery science and management, reduction of capelin, the tragedy of the commons, pollution, climate change, seals (recently protected from hunting), and foreign fishers (especially the Spanish) constituted the usual suspects in the Grand Banks debacle. But MacKenzie believes, as do many other more or less neutral observers of the fishery scene, that the mathematical models used by the scientists to depict the health of the cod stock must also absorb much of the blame: "Press the experts harder and an additional culprit emerges—the scientific models used for estimating sustainable catches. According to these models, the Grand Banks should still be full of fish. Most experts admit the models are inaccurate. . . . In the meanwhile, the models which failed the Grand Banks are being used to govern fisheries around the world."

Just before the end of the fishery, politicians on either side of the Atlantic began sniping at each other. The premier of Newfoundland, Clyde Welks, suggested that the European Community's claim of legal fishing on the Grand Banks compared favorably to Saddam Hussein's claim of legal possession of Kuwait. Other Canadian officials characterized European Union fishery officials as pirates in a fish war. In turn, European Union spokespeople accused Canadians of conducting a politicized media campaign to blame European fishers for problems created by Canada's own mismanagement.

Accusations of international interference in local fisheries are not exactly unheard of in other parts of the oceans. In 1998 the *Shen Kno*, an 80-ton Taiwanese long-liner, was caught fishing within three miles of the shore of Somalia. The skipper was fined $3 million and sentenced to amputation of his right hand and left foot. Several months later, the skipper steamed away with all of his limbs intact, but $300,000 poorer.

In 1969 the regional Grand Banks cod catch began the long downhill slide. By 1974 it was as low as 34,000 tons. In 1977 Canada extended its fishery control to 200 miles offshore, thus covering the entire Grand Banks except for two small areas called the "nose" and the "tail" of the banks. Canadian fishery scientists told the government that if appropriate catch limits were put in place, the catch should rise to 500,000 tons by the mid-1980s.

The *total allowable catch* (TAC), determined on the basis of estimates of the size of the cod stock, was set at 16 percent of the fish per year. According to the models, this size catch would allow the stock to gradually increase. In response to this good news, the government began to build up the fishing industry to prepare for the coming cod bonanza. In order to increase the economic efficiency of the fleet, tax breaks and subsidies were used to modernize existing vessels, build larger vessels, and rescue the foundering seafood companies. By the time the cod fishery collapsed, the subsidies were worth far more than the fish catch.

Almost every developed country, including the United States, did exactly the same thing when its 200-mile limit was declared. Each country viewed the declaration of offshore sovereignty as an opportunity to build up its struggling fishing industry. In the case of Canada, it was hoped that the change would revive the economies of Newfoundland and Nova Scotia, both relatively impoverished provinces. It was then that the problems began in earnest on the Grand Banks, and in most of the rest of the world's fisheries. Fisheries that were overfished by the hated foreign vessels now began to be overfished by domestic fishers. The fish population never got a rest and a chance to recover.

On the Grand Banks, the harvest by foreign ships was greatly reduced. The Spanish and Portuguese ships and those of other nationalities could still fish the nose and the tail of the banks (which harbored 5 percent to 10 percent of the total Grand Banks cod population), and also the cod banks on the nearby Flemish cap, all of which were beyond the 200-mile limit. In 1986 Canada refused to allow foreign ships to come into St. Johns, Newfoundland, for repairs and resupply, thereby adding another measure to reduce the Iberian invasion.

Despite enormous management and analysis efforts, something was going wrong. Cod stock estimates fell far short of the increases predicted by the models. The inshore fishers, which used small vessels less than 45 feet long, found that it was increasingly difficult to catch cod. They knew that the large offshore trawlers were taking too many fish, but complaints to the government, which was proud of its now prospering Grand Banks fishery, fell on deaf ears. The scientists seem to have ignored the inshore fishers as well.

The offshore fish being caught were smaller, and the fleet was catching them in a smaller and smaller area of the Grand Banks. Not to worry. It was well established, on the basis of hundreds of years of experience, that cod inexplicably disappeared from some sections of the banks from time to time. It was assumed that one of the well-known temporary shifts in fish migration patterns must be occurring.

As it turned out, the dense congregation of cod in small areas was in itself a sign of a depleted population. When only a small number of fish remain from a population that once roamed a large area, they will naturally gravitate to the best habitat, the one with the most food. When the population was large, competitive pressure for food made the fish scatter far and wide.

The estimates of the number of cod by the Canadian Department of Fisheries and Oceans (DFO), based on a random sampling survey of the banks, were much smaller than the estimates of the fishers, who did not fish randomly, but instead went to areas where the cod were congregating. One seafood company noted in 1990 that the scientific estimate of cod numbers was low because the sampling was not being done where the fish were! As far as the fishers were concerned, fishing had never been better.

During the last years of fishing leading up to the demise of the codfish, the TAC was partly based on a fish population estimate that was determined by splitting the difference between the fishers' population estimates and the estimates by DFO scientists. This calculation provided a number that made no one happy and that was indefensible scientifically. Ironically, the 1989 recommended catch of 125,000 tons was changed to 235,000 tons by fisheries minister John Crosbie, who declared the proposed 125,000-ton allowance to be so low it was "demented."

Crosbie was catastrophically wrong. In retrospect, it became clear that in the last year or two of fishing, 60 percent—not 16 percent—of the total fish stock was being removed. In January 1992 DFO scientists recommended a catch of 185,000 tons. In June 1992 they recommended that the cod fishery be closed down. Dogfish had replaced the cod

In a twist of irony, in July 2002 the same John Crosbie who had facilitated the demise of the cod fishery warned that shellfish, which replaced the cod as the main element of the Newfoundland fishing industry, were being "over fished and treated in an irresponsible manner." On the tenth anniversary of the cod collapse, Crosbie noted that provincial governments seem to have learned nothing from past mistakes. To his great credit, Crosbie seems to have learned a great deal.

Unfortunately, as the cod became more difficult to catch, some Canadian trawlers moved off the continental shelf to deep water, up to a mile in depth, to catch grenadier, hake, and eel. Fishery researchers at Memorial University in Newfoundland reported in 2006 that five of these sought-after species were now endangered and close to extinction. Fishery scientist Jennifer Devine noted that deep, cold-water species take a very long time to mature and produce fewer young than their shallow-water counterparts; hence they are very vulnerable to overfishing.

Mixing Politics and Science

The DFO was the Canadian federal agency responsible for setting the TAC on the Grand Banks. It is accurate to say that in the case of the codfish debacle this agency made one of most important and far-reaching scientific blunders of the age. But, of course, it is an agency in a democratic political system, with all that entails in terms of political involvement in decisions that are said to be scientifically based. Politicians, responding to the fishers who elect them, put pressure on the fishery administrators, who then pressure the scientists to come up with a version of the truth appropriate for the situation at hand. In this sense, science and the mathematical models were used as a cover-up, or a *fig leaf*, for irresponsible actions of the Grand Banks fishery managers.

Fishery ministers in Canada, like their politically appointed brethren everywhere, must be pliable and willing to make compromises. After all, the lives of people and the careers of politicians are at stake in the decisions they make. In addition, it seems to be a universal truth that the fishing industry will always take a short-term view of the problem and can usually be depended upon to oppose cuts in TACs. Clearly, as fisheries scientist Michael Orbach has pointed out, the study of fisheries is a combined study of politics and conservation.

For the cod fishery, as for most of earth's surface systems, whether biological or geological, the complex interaction of huge numbers of parameters made mathematical modeling on a scale of predictive accuracy that would be useful to fishers a virtual impossibility. The interaction of fish with other fish, the roles of predators and prey, the cycle of the food used by larvae and adults, the vagaries of recruitment and mortality rates, the complex food chain, the oceanographic environment in turbulent areas of the ocean where two major ocean currents mix, climactic variations, the habitat loss caused by trawlers, and many other such parameters were poorly known. Even if all of these variables were precisely understood, no one can ever know the order and intensity with which they might occur.

As a substitute for modeling the whole ecosystem, fishery scientists usually focus their models on the particular species of fish with which they are concerned at the time. When a part of the system is modeled, the assumption is made that the rest of the ecosystem will behave as "expected." Nothing unusual is expected to happen in the ecosystem, but of course it inevitably does.

An example of the single-species-focus problem—one that was used on the Grand Banks cod—is the assessment of stock size made on the basis of age profiles of the fish population. Using the age distribution of individual fish in a population of fish, the mathematical model calculates the population size that would likely be responsible for the observed ages. The model tells a fishery manager the number of fish that survived to a catchable size for a given year, and from that number it calculates the size of the cod stock. Knowing the stock size, the manager can then determine the TAC.

In arriving at the TAC, a number of assumptions have been made. One example is the age profile of a fish population (assuming you have an accurate age profile from field sampling), which is determined both by the number of fish that successfully reach maturity (recruitment) and the number of fish that die (mortality) from natural and fishing causes. But neither natural mortality nor fishing mortality is ever accurately known and, of course, each varies widely from year to year and from place to place. To get around this, a "reasonable" mortality rate is simply assumed.

James Wilson, a University of Maine fishery economist, notes that fishers have a strong distrust of government field sampling of various fish populations. There may be good reasons for this. In September 2002, the National Marine Fishery Service (NMFS) made a startling confession: Using the Research Vessel *Albatross*, NMFS had been studying fish populations for two years on the New England shelf, and all of its population size estimates were wrong. The problem was that the two trawling cables leading from the ship to the trawling net differed in length by six to eight feet. Normally, professional fishers make sure that there is no more than a two- to three-inch difference in cable length, since unequal cable lengths lead to erratic behavior of the trawling nets, including the complete closing of the net's opening in shallow water.

In an earlier test of the accuracy of the NMFS estimates, the government-owned R/V *Albatross* trawled side by side with a private fishing vessel that was using similar equipment. The catch size aboard the government vessel was about one fourth the size of that brought aboard the fisher. As it turns out, trawling is like fly-fishing in a mountain stream. There are those who catch many fish and those who catch only a few. There are lots of little tricks to hauling giant trawls (figure 1.4) that are not apparent to the untrained eye.

The mathematical models used in the assessment of the Grand Banks cod population also assumed that the size of the adult cod population had a direct bearing on the number of young fish that survived to adulthood each year. It was assumed that more adults meant more babies. The huge number of eggs produced by each spawning female cod probably provided a security blanket for belief in this assumption. Despite its intuitive correctness, the assumption was wrong because the factors in the natural environment that affect larvae are very different

Figure 1.4 A typical shrimp trawler of the type used in both Atlantic and Gulf fisheries. Problems with trawling that affect the health of local fisheries include a large bycatch (especially in the shrimp industry) that is thrown over the side. In addition, trawling tends to smooth the seafloor and reduce habitats for some of the same fish sought by the trawlers. Photo courtesy of the National Marine Fisheries Service.

from the factors that affect later stages of development. For example, much depends on whether food is available to the larvae at the moment of hatching. This is an example of a very damaging *simplifying assumption* put into a mathematical model in order to bridge a gap in understanding of the system being modeled.

Daniel Pauly and Jay Maclean, in their 2003 book *In a Perfect Ocean*, note that population estimates are further complicated by fish that are discarded at sea, fish caught but unreported, and fish caught illegally. Discards are often juveniles or fish that are caught when another species is the target. During the 1980s *high grading* was a common practice. The more valuable large cod (greater than two feet long) were kept, and previously caught smaller cod were shoveled over the side. All these practices profoundly affected the modeling of population sizes needed for fishing management and also judgments concerning the health of the marine ecosystem.

The plot is thicker than just population size. Research by Oregon State marine ecologist Mark Hinson and a number of colleagues revealed that individual fish size is also a critical factor. It is well established that big fish produce more eggs, but what these workers found in their 2004 research was that the larvae hatched from the eggs of large fish are much hardier than those derived from small fish. They refer to the big fish as *big old fat female fish*, or BOFFFs. The larvae from BOFFFs grow more rapidly and are more capable of withstanding periods of starvation than the larvae of smaller females. In addition, BOFFFs as a group have a longer time span over which they spawn relative to their smaller sisters, which maximizes the possibility of hatching at least a portion of the eggs and of having access to a good supply of food for the larvae. It is statistically more likely that small females will produce eggs at the wrong time, when the larvae find little food, and the survival rates will therefore be reduced.

Did anybody object to the Grand Banks models or the model results? In hindsight it appears that there was a storm of protest. The inshore fishermen took matters into their own hands, since their pleas to the government to stop the overfishing by the offshore trawlers went unheeded. They hired Derek Keats, a university professor from Memorial University, to evaluate DFO's published analysis of their cod numbers. The DFO fish stock numbers were off by as much as 100 percent, Keats said. In the mid-1980s one fishery expert characterized DFO's assessments of fish numbers as *non gratum anus rodentum,* or not worth a rat's ass.

The Canadian Committee on the Status of Endangered Wildlife in Canada (COSEWIC) also clashed with DFO over the fish stock numbers. Kim Bell, a fishery scientist working for COSEWIC, characterized the Atlantic cod population as a whole as endangered. Certainly an endangered species doesn't hold much potential as the basis of a fishing industry. DFO objected to this characterization.

Bill Doubleday, director of science for DFO since the mid-1980s, defended the fish number estimates of the agency by saying, "Unless you were sure you were right . . . you don't come to that conclusion [that the fish numbers were correct]. You said it was inconclusive."

Doubleday's statement says it all. It is the nature of good scientists to be skeptical, to be reserved, and to rarely assume that their numbers are absolute truths. For bureaucrats like Doubleday and politicos like fishery ministers, both under enormous pressure from every direction to keep an industry going, the constantly questioning nature of science provided the natural opening needed to ignore the warning signs of nature. Since the scientists aren't certain, why not go ahead and inflate the allowable fish catch? But Doubleday, a professional administrator of scientists, should have known better. He should have been shouting "the cod are dying" from the rooftops.

Charles Clover, in his book *The End of the Line* (2004), has few good things to say about the fishery scientists who oversaw the collapse of the Grand Banks fishery. The data on the size of the fish stocks were in the hands of a few secretive individuals, which made outside analysis of the quality of the data impossible. And when these same scientists produced the allowable catch numbers, they fell victim to the desire to bring good news to their politician bosses and to the fishing community. Good news was easy; bad news brought the roof down on them. It is a problem that plagues modeling in many specialties.

The Past and Future of Fish Modeling

Basing allowable catches on *maximum sustainable yield* (MSY) was the goal of DFO's hugely unsuccessful management of the cod on the Grand Banks of Newfoundland. Through the use of mathematical models, a level of fishing that could sustain itself indefinitely was sought. The concept of MSY is based on the assumptions that any species in the sea will each year produce a harvestable surplus and that if you take back to the dock that much and no more, you can keep harvesting it forever.

The concept of MSY was introduced in the late nineteenth century but reached its heyday in the 1930s. It was the governing concept of fishery science during the 1940s and 1950s and still finds widespread use in applied fishery science, although it sometimes appears under a pseudonym. The scientists who supported MSY believed that fish were the great integrators of the environment, the peak of the food chain, and they ignored the rest of the ecosystem.

MSY was a commonsense concept that brushed aside the old *doctrine of traditional limnology*, which took a more holistic view of marine ecology. These traditional scientists assumed that the fish were part of a living community within a larger ecosystem. Fish species interacted with one another and with other organisms, both plants and animals, in a very intricate and balanced process of feeding, growing, reproducing, and dying. The water circulation, storms, bottom sediment, and bottom shape all played some role. It was all very complex, too complex to model and come up with useful answers. Then along came MSY, and all this immensely complex system could be mathematically bypassed.

In the end, it *was* too good to be true. Doubt began to creep into the minds of even the most loyal MSY supporters starting in the 1960s.

The practical application of the models was proving impossible, and fish stocks declined. As far as scientists are concerned, the end of MSY came in 1977, when P. A. Larkin, a Canadian fisheries biologist, wrote his famous marine fisheries paper titled "An Epitaph for the Concept of Maximum Sustainable Yield." He ended the article with a genuine epitaph:

> MSY
> 1930's–1970's
>
> Here lies the concept MSY
> It advocated yields too high
> And didn't spell out how to slice the pie
> We bury it with the best of wishes
> Especially on behalf of fishes
> We don't know yet what will take its place
> But hope it's as good for the human race.

Larkin pointed out four major shortcomings of MSY:

• **Fishing to the MSY creates the possibility of population collapse**. If one fishes right to the limits of the MSY for a species, most of the fish that are caught will be young and first-time spawners, because this class of fish is the healthiest and has the lowest mortality rate. The problem is that first-time spawners don't produce the best eggs; they're not as good as BOFFFs. In addition, if only one age group is spawning, a failure in egg hatching or larval survival could lead to a calamitous population failure. Putting it another way, a population of fish that is being harvested at maximum sustainable yield is much more unstable than an unfished population. Once a population collapses, it may stay low for a long time, such as the cod or haddock on the Grand Banks.

• **Fishing to the MSY may reduce genetic variability in a fish population**. Most fishes can be divided into subpopulations, all of which may have different ideal MSYs. Salmon subpopulations spawn in different rivers. Cod subpopulations, of which at least a dozen are recognized, migrate to different shallow-water areas during the summer. Some of the subpopulations recover from fishing pressure more slowly than others. Consequently, if a population is fished at or near the MSY, subpopulations that reproduce the slowest will be hit hardest. The danger is that the remaining fish available for catch will all be from one subpopulation. Under ideal circumstances, in order to preserve essential genetic variability within a healthy fish population, the MSY should be based on the harvestability of the most vulnerable subpopulation.

• **The MSY does not accommodate the interactions among the species of organisms that constitute an entire aquatic community**. Fish species have very complex interrelationships, the classic example being the cod-mackerel-herring saga. Herring eat cod eggs and mackerel eat herring. If the mackerel are depleted by fishing, herring become more abundant and eat more cod eggs, to the detriment of the cod population. Another example is the growing Atlantic squid fishery that has resulted in an important and possibly damaging reduction of food supply for porpoises. The doctrine of traditional limnology was right; one species should not be managed out of context with other fish species.

• **The MSY concept is completely irrelevant for recreational fishers** (figure 1.5).

Larkin's paper and poetry came too late. The simple and appealing concept of MSY was already deeply entrenched in the world of fishing politics. The fishery scientists who believed in MSY had been too successful in selling the idea.

The U.S. fisheries are governed by a related concept called *optimum yield* or *optimum social yield*, as defined by the Magnuson-Stevens Fishery Conservation and Management Act of 1976: "Conservation and management measures shall prevent over fishing while delivering optimum yield from each fishery on a continuing basis. Optimum yield is the maximum sustainable yield modified by any relevant economic, social or ecological factors."

The term "over fishing" is not defined in the Magnuson-Stevens Act, and the definition of optimum yield is so vague that it could justify any level of fishing, even one exceeding the MSY. At least the act recognizes that the industry is also made up of people and not just fish. Modeling the aquatic ecosystem is hugely complex, and throwing in economic and social factors as well only increases complexity. However, recent modifications to the Magnuson-Stevens Act require not only that fishing pressure be reduced on an overfished resource but also that at the same time the stock must increase.

Today, no matter how it is phrased, governments keep asking scientists for advice on catch limits, or the MSY. Basically, our politicians have decided that catch control is the best and easiest way to control fisheries. If that is what government wants and pays its researchers to study, then that is what we will get, despite the fact that it is increasingly obvious

Figure 1.5 Recreational fishing in the surf zone off Cape Hatteras, North Carolina. The impact of recreational fishing on fish populations is often difficult, if not impossible, to determine. Photo courtesy of the National Atmospheric and Oceanic Administration.

that the huge complexities in "fish population science" extend far beyond simply establishing catch limits.

Some fishery scientists argue that it would be easier in a practical sense to use an effort-control approach. Fishing effort would be regulated by controlling the size of the boats, the size of the fishing area, and the dates within which fishing can occur. An example of this approach is the oyster fishery in Chesapeake Bay, which requires using hand-operated dredges from small skipjack sailing vessels. But there doesn't appear to be much difference between catch control and effort control. Both must adhere to some sort of sustained yield principle if the fishery is to survive. As it turns out, the oyster fishery in Chesapeake Bay is in catastrophic decline, as illustrated by a reduction in harvest from millions of bushels annually in the 1880s to 3,800 bushels in 1999.

Still another management method is the establishment of no-fishing zones, which allow stock recovery, especially among the BOFFFs, which is necessary to reestablish a healthy fish population.

Fisheries in the United States are managed by eight regional fishery management councils, made up mostly of politically appointed nonscientists who represent various interests (e.g., commercial fishing, recreational fishing, tourism, and state government). As is so often the case with citizen councils, individuals on the council view themselves as lobbyists for their particular constituency. In 2003 and 2004, respectively, the Pew Charitable Trust and the U.S. Commission on Ocean Policy agreed that the regional management councils will not be able to solve the overfishing problem. The problem, as stated in a Pew Ocean Science Series report, is that "most council members [around 80 percent] are affiliated with or reflect commercial and recreational fishing interests. Virtually none comes from the conservation world or the public at large."

When evidence that a fishery is in danger is presented, acquired from the results of a recognized model analysis, regional management councils are more likely to respond than if raw numbers are placed before them. NMFS scientists argue that since models are alleged to provide a long view, this is one way to jar the councils off a very short-term, economically driven view of a fish stock. Perhaps it also has something to do with the mystique of models and the mystery and apparent sophistication of mathematics.

An example of a recent model "victory" is the use of the *virtual population analysis* (VPA) model to convince the South Atlantic Regional Fishery Management Council of the need to regulate and eventually close the red porgy fishery. The red porgy is both a commercial and a recreational fish whose numbers and size are in decline. Recruitment is down 95 percent.

VPA requires knowledge of the number of fish caught from a single-year class each year for several years, as well as the mortality rate from fishing and natural causes. All of these numbers are difficult to come by. Mortality rates of a population under extreme stress, as in the case of the red porgy, are especially vague. Whether right or wrong, the mathematical model was used to demonstrate what was already quite obvious from catch and fish size numbers. But in this case, a precise or quantitative prediction should have been neither needed nor expected—the qualitative indication that the species was in deep trouble should have sufficed. In spite of the fact that the red porgy is believed to be the most overfished species in the South Atlantic and in spite of the VPA analysis, it still took five years of debate on the council to limit fishing of the species. And even then, the North Carolina representative objected to the decision.

Certainly it would be difficult to claim that the use of mathematical models in fishery science has resulted in a stable and healthy world fishery. In fact, the world's fisheries are deteriorating, and mathematical modeling efforts are reaching a peak. Two-thirds of the marine stocks in the Atlantic and Pacific oceans (including cod, shark, lobster, and shrimp) are either gone, overfished, in strong decline, or being exploited to the maximum extent possible. But models aren't the only villains. How much of this deteriorating situation is attributable to models and how much to politics is difficult to determine. In developing countries, chaos often substitutes for management, and even perfectly functioning mathematical models would make little difference.

There are some positive signs indicating that at least in some fishery research scientist circles, mostly in the academy, *"what if" modeling* has arrived to act as a guide to fishery management. Such qualitative models are used to seek general guidance for management and not specific defined numbers such as TAC or MSY. "What if" modeling provides one way of evaluating alternative approaches to solving a fishery problem. What if the fishery is closed? What if fishing is permitted only on a certain area of the continental shelf or during a certain season? What if mesh size of nets is increased? Decreased? In this kind of model it's not necessary that all the assumptions behind the model be completely understood or that all the parameters that affect the ecosystem be included, so long as the most important ones are taken into account. General questions are asked and general answers are received. A high degree of accuracy is not expected.

Another approach that can bypass or minimize the use of mathematical models altogether, one that seems to be particularly favored by those concerned with U.S. Pacific fisheries, is *adaptive management*— essentially "Give it a try and adapt as time goes by." This is particularly suited for the management of marine reserves or no-fishing zones. For example, once a marine reserve is designated, study it and see if BOFFFs increase. Observe what this does for the population that is being fished outside of the reserve. Adaptive management approaches could involve moving, enlarging, or shrinking the marine reserve, or even making it a seasonal reserve or a reserve for certain species only. The marine reserve approach may be the only way to encourage increases in the numbers of BOFFFs.

Most fishery scientists seem to recognize that Larkin was right in his criticisms of MSY, and there is widespread agreement that accurate mathematical modeling of the complex marine ecosystem for fishery purposes is probably impossible. Yet mathematical modeling to come up with an allowable catch number continues to be the mainstay of fishery management. We believe the quantitative mathematical models actually used in fishery management fall into two categories.

Category 1: Modeling Blindfolded. Fishery modeling is done by nonbiologists or those biologists who are deeply ensconced in the political system, where their hand is forced in the direction of finding the politically acceptable and most optimistic answer. It is a fact of life that the basic researchers who formulate the models are usually not the ones who actually apply them in the chaotic tangle of special interests in a democratic society. The models are often applied and released to the public without explanation or discussion of the uncertainties.

Such blind or rote application of models, whatever the reason, is the problem addressed by Raymond Beverton and Sidney Holt, authors of a once widely used mathematical model that bears their names, using fish age profiles to calculate the population of a fish species. Both expressed concern about the simplistic way that their model and other models are actually applied by fish managers. According to Beverton, "There is a strong inverse relationship between the growth of fisheries science and the effectiveness with which it is applied." In Deborah MacKenzie's *New Scientist* article, Holt is quoted as noting: "It has been extremely difficult to dissuade fisheries biologists from applying simple formulas like recipes and getting half-baked answers."

Category 2: Models as Fig Leaves, Shields, and Clubs. Peter Aldrich, an NMFS modeling expert and model realist, argues nonetheless that models have proved to be very useful because of their value in winning converts for reducing catch levels to save a fish species. Models have a reputation as the state-of-the-art, sophisticated approach to solving the problem of the dying American fishing industry. They give all interests something to hang their hats on, something to use by way of explanation to disappointed constituents, something to hide behind, something to use as a club. Certainly, the argument goes, the use of a mathematical model to reduce fishing pressure on a species, even if the model is wrong, is better than the alternative of having to sort through some tabulated raw field data accompanied by the opinion of an "expert," only to be refuted by the opinion of another "expert."

Models can also serve as strong insulators, protecting agency scientists and fishery managers from direct attack by politicians who are anxious to please the unhappy fishers among their constituents.

As we move into the twenty-first century, we are not even close to accurate quantitative modeling of any significant portion of the marine ecosystem. Experience in regulating fishing and the catastrophe of the Grand Banks are probably getting us closer to good fish management, but it's questionable whether the knowledge gained from quantitative mathematical models is helping in that regard. Single-species models can't work and protect the entire ecosystem, but single-species models are really all we use. Just like the studies at Yucca Mountain (chapter 3), it seems as though the more we know about fisheries, the less we know. Each step in the direction of understanding ecosystems reveals more and more complexities, and in any complex system in nature we can never obtain quantitative modeling answers at the level of accuracy that society needs.

Society seeks an answer through fish mathematical models, but it can never get that answer from them. Turning away from the fishery models, however, may be akin to reversing a high-speed freight train that's rolling downhill. It won't be easy.

Prediction is very difficult, especially if it's about the future.

—*Neils Bohr, Nobel Prize–winning physicist*

chapter two

mathematical models

escaping from reality

War by the Numbers

During World War II, military mathematical modeling, or *operational research*, became a critical tool for analyzing the war experience. One of the more successful applications of mathematical models resulted in a large increase in the sinking of U-boats by the British navy, after studies suggested new tactics and new settings for depth charges. Operational research was also responsible for suggesting that large convoys of merchant ships were safer than small convoys, the opposite of contemporary military thinking on the subject.

However, the low point of the military mathematical model may have come and gone during the Vietnam War, when modeling the battlefield proved difficult and disastrous. Robert McNamara, one of the ten Ford Motor Company whiz kids, instituted a numbers-only mentality in the management of everything from industry to the World Bank. It may have worked quite well at Ford where the whiz kids, hired by Henry Ford II, turned around a foundering company. But this mentality applied to war was another matter.

McNamara, today best known for the fiasco he helped to create in Vietnam, emphasized numbers, costs, and efficiency, while downplaying the role of human intuition. Once when a White House aide said that the war was doomed to failure, McNamara reportedly responded: "Where is your data? Give me something I can put in a computer. Don't give me your poetry." Twenty years after the war was over, Mr. McNamara admitted that his approach to managing the war was "terribly wrong."

Allain Enthoven, now a chaired professor at Stanford University, was the chief whiz kid and systems analyst for McNamara. The McNamara and Enthoven approach to managing war was cold as a fish, quantitative, impersonal, objective, and lacking in intuition and common sense. Events proved that these *rational modelers* had a fatal flaw: they were unable to admit failure.

One infamous part of quantitative warfare in Vietnam was the notoriously inaccurate enemy body count, considered a measure of success in the war. The body count for remote air and artillery strikes was mathematically modeled to determine how many people would be killed by a certain tonnage and type of explosives and the number of napalm canisters, taking into account the terrain, the vegetation, and the density of people on the ground, among other factors. In closer combat involving infantry units, individual combatants tracked the body count. It was a mathematical model vulnerable to manipulation because evaluations, promotions, commendations, and decorations for officers and noncoms were at stake, depending on the results.

As we have already seen, models may be far from objective when human choices and politics play a part in the process. Arriving at high body counts in Vietnam perhaps was easier than going against the grain with more accurate counts, just as going against the grain of an assumed robust cod population on the Grand Banks by reporting more realistic figures proved difficult for fisheries managers. Eventually all dead bodies became enemy dead bodies—"If it's dead and Vietnamese, it's VC." was the gruesome saying of the times.

The body-count modeling problems are obvious, especially in hindsight. They provide important lessons for all quantitative mathematical modeling.

- *Political objectives polluted the models.* The perception that the war was being won was important in order to sustain support back home.

- *The wrong question was asked.* The body count was not a good measure of success of the American army against a highly motivated, disciplined peasant army.
- *No one looked back.* The veracity of the modeling effort should have been confirmed by field checks.

Models, Models, and Models

A *mathematical model* is a description of a process or a prediction about the end result of a process, expressed as an equation or equations. A model is a numerical analogue—a set of equations that describes the relationships between parameters that control a process. In this book we talk mostly about mathematical models that are said to describe or predict with useful accuracy something about large-scale processes on the surface of the earth. This includes both physical and biological processes. All of the model dominions in this book are *applied models*, or societal relevant models used for engineering, policy, financial, or management purposes.

Quantitative mathematical models are predictive models that answer the questions *where, when,* and *how much.* Where will the invasive plant spread next? When will an artificial beach disappear? How much will the sea level rise in the next century?

By contrast, *qualitative mathematical models* are used to predict directions and magnitudes. For example, is a plant likely to be invasive? Will sea level rise or fall? Will the available fish for harvest be large or small? Will the global climate warm or cool? These models also seek the answer to the questions *why, how,* and *what if.* Why is plant species X invasive, while plant species Y is not? How will the nourished beach disappear, and what mechanisms are likely to be responsible for beach loss? What if trawling for a fish species is halted and only long-lining for the species is allowed? What if rainfall increases at Yucca Mountain's proposed nuclear waste storage site?

The distinction between quantitative and qualitative models is a critical one. The principal message in this volume is that quantitative models predicting the outcome of natural processes on the surface of the earth don't work. On the other hand, qualitative models, when applied correctly, can be valuable tools for understanding those processes.

There are a number of other categories of models as well, sometimes rather vaguely defined. *Statistical models* are those based on statistical studies of past events for the purpose of estimating the probabilistic future behavior of the system. This type of modeling is often used in the social and health sciences. The insurance industry, for example, determines premiums based on statistical models of health data. *Simulations* mimic an event to determine what might transpire. For example, hurricanes, floods, and landslides are often simulated, as are nuclear weapon explosions, battles, and damage to spacecraft in orbit.

Quantitative models may be categorized as either analytical or numerical. *Analytical models* involve simple equations that can be solved rather readily, perhaps using only paper and pencil. *Numerical models* are much more complex, may involve differential equations, and are often solved with complex computer codes.

Determining *model sensitivity* is a method used to resolve the relative importance of the various factors that make a process work. Various components of the equations are changed to see if the outcome of the model changes. Is wave height more important than wave angle or grain size of the sand on a beach in determining sand transport in the surf zone? An important parameter will make a big difference in the final answer and an unimportant factor won't make much difference. *It is important, however, to recognize that the sensitivity of the parameter in the equation is what is being determined, not the sensitivity of the parameter in nature.* If the model is well founded, determining the sensitivity of various parameters is a valid exercise. If the model is wrong or if it is a poor representation of reality, determining the sensitivity of an individual parameter in the model is a meaningless pursuit.

Another distinction between qualitative and quantitative models is the kind of answer that a model provides. If the answer is a single number, the model is quantitative. For example, a quantitative model might predict that the global atmospheric temperature will rise by 3 degrees Centigrade, plus or minus 1 degree, over the next century, whereas a qualitative model might predict that the temperature will continue to increase over the next century, with a possibility that the rate of temperature rise will accelerate. In another example, quantitative modeling is the prediction that because of sea-level rise, the shoreline will retreat 170 feet, plus or minus 30 feet, over the next century. The qualitative equivalent might be a prediction that the shoreline will continue to retreat and probably the rate of retreat will accelerate over the next century. Whether the path to an answer is analytical or numerical, a quantitative answer comes from a quantitative model. The same goes for qualitative models.

In a qualitative model, because one is determining only the direction of a process or the basic mechanics behind a process, only the most important variables need to be considered. Because of the omission of minor processes, the results of all qualitative models may be imprecise or wrong to some degree, but that does not matter so long as the qualitative question at hand can be reasonably answered. Quantitative models require a great deal more accuracy, and to make an accurate prediction a process must be completely understood. All variables of any importance, including feedbacks, must be accounted for if the model is to answer the question at hand.

The actual model may be expressed in one or several relatively simple equations (see appendix), but the calculations using these equations that apply to a large area of the earth's surface through time may be very complex. The method of calculation required for the application of a model is known as the *computer code.*

A single computer simulation of a natural process over time and space may involve hundreds of lines of equations. Imagine the fifteen-year effort involving a small army of specialists that Microsoft went through to develop the word-processing program used in typing the drafts of this page. Millions of dollars were spent in debugging Microsoft Word, yet as anyone who uses a word processor knows, bugs still exist, albeit mostly very minor ones. Programs behind the models that we discuss in this volume have for the most part not been through a detailed quality assurance program. So the question always exists: does the software or computer code actually model what the authors say it models? Programmers know that inevitably there will be many bugs; the hope is that they will all be minor ones.

In chapter 3 we deal with a complex super model, actually based on hundreds of models, to predict the fate of nuclear waste stored at Yucca Mountain, Nevada. These computer codes must describe hundreds of physical, biological, and chemical events that occur over long periods of time over a wide area of the earth's surface. The potential for computer code error is vast, and it is very difficult to evaluate.

A good modeling approach is to "*open-source*" the codes for any and all who are interested. In a recent controversy concerning the shape of the global warming curve over time, however, the scientists who came up with the curve refused to allow others to inspect their computer code. As a result, a pall of suspicion has fallen over their results.

Equally crucial is providing a list of all important assumptions behind models—but this can be tricky. For example, one might say that the assumed average wave height in a mathematical model to predict sand transport is six feet. But the story behind that assumption is more complex. To fully understand the average wave height number, one must accept the following sub-assumptions:

- All waves come from the same direction.
- All waves are of the same height.
- Future wave conditions will be the same as those in the past.

Naomi Oreskes, science historian and modeling philosopher of the University of California at San Diego, uses Lord Kelvin to provide an illustration of the hazards created in earlier times by the drive for quantification. William Thomson, otherwise known as Lord Kelvin of Kelvin temperature scale fame (figure 2.1), was one of the leading physicists of the latter half of the nineteenth century. More than 100 years ago, in Lord Kelvin's time, there was much uncertainty about the earth's age. This was before the onset of techniques to determine ages by rates of decay of radioactive elements. Estimates by geologists ranged from 100 million years to hundreds of billions of years, but most geologists, more or less correctly, thought that the age must be on the order of a few billion years. Current thinking is 4.5 billion years. To come to their conclusions, the geologists used a *conceptual model* based on observation, past history, and experience, spiced with a dose of intuition. A conceptual model is a qualitative one in which the description or prediction can be expressed as written or spoken words or by technical drawings or even cartoons. The model provides an explanation for how something works—the rules behind some process.

The conceptual model that provided a qualitative age estimate was based on the *Principle of Uniformitarianism*, which holds that the present is the key to the past. It is assumed that the processes that mold and shape the surface of the earth today must have worked the same way in the past. Judging from the rate at which streams, blowing wind, and glaciers remove and deposit sediment today, and the frequency of volcanic eruptions, the geologists calculated, in extremely rough form, that it must have taken billions of years for the earth to come to its present state.

Lord Kelvin, unconvinced by such a crude approach, obtained an age of 98 million years, on the assumption that the earth had started out as a molten body and had been cooling ever since. This determination could be shown using a simple mathematical model, which could be calculated by hand. Kelvin's method was a quantitative and precise way to get at the

Figure 2.1 Physicist William Thomson, otherwise known as Lord Kelvin, estimated the age of the earth to be 93 million years, on the assumption that the planet began in a molten state. The simple model he used to calculate the age was valid, but the underlying assumption was wrong. Geologists using conceptual models correctly determined that the planet was much older, but Lord Kelvin's age estimate remained credible (until the role of radioactive elements was discovered) for a few years, an early example of the quantitative trumping the qualitative. Photo from answers.com.

earth's age. And since its basis was a principle of physics (cooling rate), the results were widely accepted. Lord Kelvin declared much of geologic thinking about fossils, stratigraphy, and earth history to be invalid. He also cast doubt on Darwin's theory of evolution because, according to his concept, the earth had been at its current temperature for only a short time span, too short for evolution to operate.

Alongside the shaky qualitative conceptual models of the struggling field geologists of his day, Lord Kelvin's number, derived from a valid mathematical model, seemed to be a precise and reproducible thing of beauty. Combined with his forceful personality, Lord Kelvin's declaration plunged geology into a virtual dark age that held back progress in both earth science and evolutionary theory for a few years.

But Lord Kelvin was wrong, and it was the discovery of the continuous production of heat by the decay of radioactive elements in the earth's upper layers that finally countered his idea. The present temperature of the earth was not derived from the cooling down of a molten body. Instead, because heat is generated in the crust by radioactive decay of a number of elements, including uranium, the earth has steadily maintained its current temperature for a very long time. Otherwise, we would be looking toward a very cold earth on the not-too-distant horizon. Interestingly, Lord Kelvin's age of the earth is still supported by a number of creationists in their battles with modern earth science.

Lord Kelvin's model was an early example of a quantitative model trumping a qualitative one, a common problem even today. His model of the rate of cooling was perfectly valid—that is, the principles of physics he applied were correct. The cooling of the earth is not a complex process, and a quantitative model can successfully describe it. His mistakes were the underlying assumption of a molten beginning of the earth and the failure to understand the importance of radioactive decay as a source of heat in the earth's crust. The important lesson here is that no model can overcome a series of bad assumptions.

In hindsight, it is hard to see a way that Lord Kelvin could have guessed the truth. His was a *situational bias*, the phenomenon by which our thinking is so obscured by our present state of knowledge and known conditions and observed trends that we are blinded to the future. It is hard to get out of one's own cocoon.

Still another conceptual model of the age of the earth dominated Western thought for more than 1,000 years. It was the biblical chronology model, which began with the publication of *Chronologia* in A.D. 212 by a priest and former Roman soldier named Julius Africanus. The "chronologists" were trying to determine the date of the Second Coming of Christ, in order to understand when the thousand-year reign revealed in the book of Revelation would begin. The widely held assumption at the time was that the Second Coming would occur 6,000 years after the earth was formed. Thus the age of the earth was needed in order to determine the date. The 6,000-year assumption was based on two sources of information. The first was Elijah's prophecy in the Jewish Talmud that the earth would last 6,000 years. The second was the belief that each day in the seven days of creation described in Genesis was in reality 1,000 years and that Christ would return for the seventh day of rest.

Africanus totaled "known" time spans of biblical lives and events, starting with Adam. The Septuagint version of the Hebrew Bible was the source of the data on the early part of the earth's existence, and it revealed that Adam lived for 930 years, Noah for 950, Moses for 120, and Abraham for 175. Adding up all the life spans, Africanus concluded that Christ was born 5,500 years after the formation of the earth and that he would return in the year A.D. 500.

Subsequent chronologists, including Martin Luther, adjusted the date of the Second Coming by a process we would now call *model tweaking*. According to Jack Repcheck's fascinating account of this in *The Man Who Found Time*, "the chronologists [that followed Africanus] were consistent in putting off [the Second Coming] until a couple of hundred years after their own deaths." As is often the case in some modern modeling endeavors, so much uncertainty existed in the original numbers that tweaking was carried out without raising questions of credibility.

The last and perhaps most famous chronologist was James Ussher, the Calvinist archbishop of Armaugh (Ireland), who pronounced in a 2,000-page book published in 1650 that creation of the earth started at noon on Sunday, October 23, 4004 B.C. By his reckoning, Christ should have returned around October 23, 1996.

The age of the earth according to geologists is much greater than Ussher's reckoning. The conceptual model of the chronologists failed for a number of reasons. Like Lord Kelvin's model of a cooling earth, the methodology of the model was reasonable enough, but the underlying assumption was unsound. Counting biblical generations and events is a valid approach (assuming that everything was recorded accurately in the Bible), but the assumption that Adam came along when the earth began has no basis in science.

Faith-based assumptions and conceptual modeling are clearly immiscible. But we will demonstrate that applied mathematical modeling is at times no less biased, skewed, or slanted by political correctness, advocacy, or economic interests than the biblical slant of the chronologists.

Fast-forwarding to the late twentieth century, we confront another celebrated failure of quantitative modeling. It began with the 1972 publication of *Limits to Growth*, a book commissioned by the Club of Rome. The club, a secretive think tank started by a distinguished British research chemist and a successful Italian industrialist in 1966, today consists of around a hundred economists, businesspeople, scientists, and government officials from fifty-two countries on six continents. The club's book famously predicted that within the coming hundred years, there would be widespread natural resource shortages and economic collapses. The authors warned that unless immediate action was taken to control population and pollution, we would not be able to turn the situation around. This doomsday prediction was based on a mathematical model known as the *pessimist model*. Unlike the simple analytical model applied by Lord Kelvin, this was a more complex model called World III and requiring extensive computer calculations. The document argued that population growth and pollution from industrial expansion were leading to total exhaustion of natural resources and massive environmental destruction. It predicted that catastrophes would begin by the year 2000.

There were many problems with the model. It treated the earth's mineral reserves as fixed and unchanging. This decidedly static view of economics and unhistorical understanding of human creativity held that we would run out of oil according to a time schedule calculated from what was then known about reserves and production methods. It ignored the possibility of additional major oil discoveries, advances in petroleum exploration and extraction technology, and the possible contributions of nuclear, solar, or wind energy sources. The model also assumed that food production per unit of land area would remain steady.

Oreskes notes: "In effect [earth] scientists treat the systems they are modeling as though the systems were static. This is not to say that the modelers believe the systems are static—no earth scientist could imagine any system as truly static. Nonetheless scientists often imbed stasis into their models."

University of Manitoba professor Vaclav Smil summed up his view by noting that the *Limits to Growth* report "pretended to capture the intricate [global] interactions of population, economy, natural resources, industrial production and environmental pollution with less than 150 lines of simple equations using dubious assumptions to tie together sweeping categories of meaningless variables."

The problems with the model went beyond the huge technical weaknesses. It was an example of an *advocacy model*. A Club of Rome official stated shortly after the predictions were released that the idea was "to get a message across, and to make people aware of the impending crisis." In other words, the model outcome had been determined before the model was run. Finding the truth according to a preconceived opinion or philosophy is a common flaw in applied mathematical modeling. And it is very similar to finding truth that matches one's religious faith.

The *optimist model* emerged in a 1976 book titled *The Next 200 Years*, by Herman Kahn and others. This volume presented a view of the future that could be briefly stated as "necessity is the mother of all invention." Kahn basically argued that when the need for more food arises, better technology will save the day. When the price of oil soars out of sight, other sources of energy will come to the fore. This model is a qualitative conceptual model, based simply on a number of scenarios devised by the authors.

Both the pessimist model and the optimist model were derived from the same database. The difference is in the assumptions made and in the personal views of the modelers. *Personal view models* are those that are slanted to prove the belief of the modeler.

Ideally, comparing model results with a real-world situation, a process known as *calibration or validation*, tests a model. That is, an attempt is made to "predict" an event that has already occurred using the model in question. For example one could *hindcast* the cod failure on the Grand Banks.

However, one successful calibration or one successful prediction does not mean the next attempt at calibration will also pass muster. As Naomi Oreskes argues, successful reproduction of an event in a complex natural system is no guarantee that the model will accurately predict or describe the next such event. In fact, she argues that most likely it won't make a successful subsequent prediction. Calibration may show that a model fails to reproduce a situation, but the converse is not always true. Leonard Konikow and John Bredehoeft, geologists with the U.S. Geological Survey, made the same point in a famous 1992 paper titled "Ground-Water Models Cannot Be Validated." The Konikow and Bredehoeft paper received the Meinzer award from the Geological Society of America, but their paper and the views of Oreskes seem to have had minimal impact in the modeling community. Model calibration and validation are alive and well.

In some types of modeling, a second calibration, known as verification, is carried out. The model is first calibrated with one set of events and then verified with a second set. It could work like this: The model is tweaked so it successfully predicts shoreline erosion along a stretch of coast that occurred between 1950 and 1970. The tweaked model is then verified by application to the known erosion rate between 1970 and 1990. If it successfully predicts the erosion rate between 1970 and 1990, the model is said to be verified and can be used to predict the future.

Perhaps the single most important reason that quantitative predictive mathematical models of natural processes on the earth don't work and can't work has to do with *ordering complexity*. Interactions among the numerous components of a complex system occur in unpredictable and unexpected sequences. In a complex natural process, the various parameters that run it may kick in at various times, intensities, and directions, or they may operate for various time spans. Chapters 5 and 6 provide examples of this phenomenon, with lists of dozens of parameters that may affect the natural processes of shoreline erosion and longshore transport of beach sand, respectively. Parameter after parameter kicks in and out—who knows when, where, and for how long. Complicating things even more are positive and negative feedbacks.

Complexity is a big thing in today's modeling world. There are circulating newsletters, books, technical journals, societies, scientific meetings, and branches of funding agencies that are concerned almost exclusively with complexity. The term is formally defined in a number of (complex) ways, but we will stick with the (relatively) simple explanatory description in the previous paragraph.

William Sherden is a marketing consultant, a Stanford University professor, and the author of *The Fortune Sellers*, a book that provides a skeptical view of stock market forecasting. He notes that complex systems are so highly interconnected with numerous positive and negative feedback loops that they often have counterintuitive cause-and-effect results, as when the "addition of a new highway to alleviate a traffic jam causes the traffic jam to become worse," a *negative feedback*. The rich getting richer and the poor getting poorer are both examples of *positive feedbacks*.

Global warming could lead to melting of the Arctic Ocean ice cover, leading to increased evaporation of ocean water, leading to more precipitation in the Arctic region. Increased snowfall leads to increased accumulation of ice leads to a new ice age. Thus, global warming leads to global cooling, a negative feedback of global proportions.

One reason why earth systems are complex has to do with the relationships between the variables that make a system work. A *linear relationship* is one in which variables increase or decrease at a uniform rate—a straight line on a graph. Most relationships between parameters in a complex earth surface process, however, are *nonlinear relationships*. As one variable changes, another may change exponentially. What complicates the relationships between the numerous parameters that control any natural process even more is the fact that a number of them may change simultaneously as a natural process unfolds. A relationship that may be linear in isolation may be nonlinear in the context of simultaneous changes in other parameters. The reality of any natural process on the earth's surface is a convoluted bird's nest of interrelationships. Complexity reigns, and that is the beauty of the natural world. An example is Yucca Mountain (chapter 3), where as the downward rate of water flow increases through the rocks, the volume of water transported increases disproportionately. This is a positive feedback and a nonlinear relationship.

In classic physics, by contrast, the systems being dealt with are usually not complex, in the sense used here. Modeling in physics is labeled *determinism*, in that prediction of events is possible. Thus we are successful in prediction of the future positions of the planets, the times and dates of eclipses, the rate at which radioactive elements decay, and the time it will take a ball to roll down an inclined plane.

The *New York Times* on June 7, 2004, noted: "In New York City sunrise will be at 5:25 a.m. Eastern time on Tuesday, and Venus is to begin leaving the solar disc at 7:06 a.m., when the sun is 17 degrees above the horizon. The planet's final contact with the sun's edge should occur about 7:26 a.m. when the sun is 20 degrees high. There will be another transit on June 6, 2012. After that, the next ones will occur in 2117 and 2125." What a contrast to prediction of events in complex systems like beaches, global climate, fisheries, rivers, the stock market, and invasive plants!

The same predictive success is possible in the engineering design of bridges and elevated water tanks. The laws of physics apply well to steel and concrete. Plus, humanity has accrued a great deal of experience with these materials to sharpen predictions.

Engineering design and prediction always have a large *safety factor* to allow for human error and to assure that structural safety predictions will be right. Designs are intended to last a certain length of time, to withstand a certain wind velocity or an earthquake of a specified magnitude.

Modeling in any system that results in a single answer (right or wrong) without any indication of the possible range of error in the answer is a *deterministic approach*. In most applied quantitative modeling of earth processes, the results should be *probabilistic* to express the uncertainties involved. That is, the answers should have an error bar or a plus or minus expression of the possible range within which the correct answer must lie.

It is a catch-22 situation. Modelers view error bars as a valid response to critics of quantitative mathematical models, but you can't determine accurate error bars for a prediction without having the same level of model accuracy that is needed to get accurate deterministic answers. Furthermore, an invalid model doesn't provide a valid answer, whether you use error bars or not.

The *error envelope*, or *cone of uncertainty*, on a predicted hurricane path (figure 2.2) is an example of a very useful error bar for quantitative mathematical models. The National Hurricane Center and the Weather Channel produce maps showing an ever-widening funnel of possible storm impact areas in the direction of storm movement. The funnel is centered on the most likely predicted line of storm movement for as

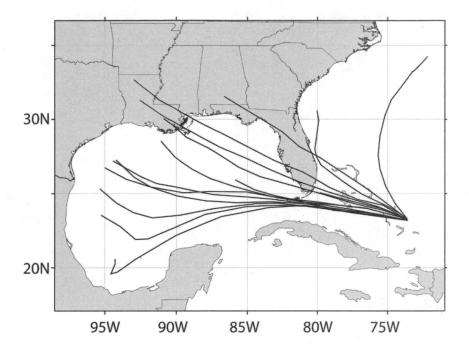

Figure 2.2 Modeled hurricane tracks for Hurricane Rita in 2005. The tracks form a cone of uncertainty, which, while frustrating to coastal dwellers, is a straightforward way to represent the uncertainties of hurricane track predictions. Diagram from Colorado State University Department of Atmospheric Sciences.

much as three days in advance. Most people in hurricane-prone areas probably have an intuitive feeling about the accuracy of hurricane model predictions because in the past many have stocked up with extra food supplies and/or evacuated their homes only to find that the sun is shining and a gentle breeze is blowing on the predicted day of storm arrival.

Meteorologists are up front about the uncertainties of their hurricane path predictions, which are high, even though the models have an excellent statistical or experience base. There must have been much gnashing of teeth at the Hurricane Center when Hurricane Dennis (1999), located off Cape Hatteras, halted and then reversed its path and began moving south, a most unusual path. Teeth gnashing of an even higher order must have occurred when Hurricane Ivan (2004) passed across the Gulf of Mexico shoreline, through the state of Virginia, and then made a wide southerly arc out into the atmosphere over the Atlantic Ocean. Eventually the remnants of Ivan returned to the Gulf of Mexico and crossed the Gulf shoreline for a second time!

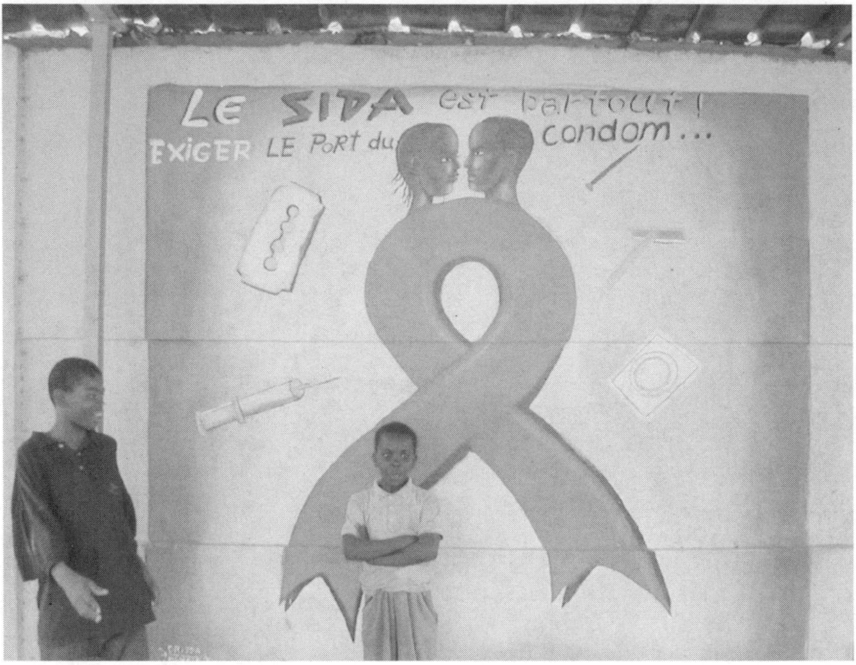

Figure 2.3 AIDS education poster in Mali, Africa. Countries in Africa with extensive education campaigns are having some success in holding down HIV numbers. Mathematical modeling by the United Nations of the extent of this societal catastrophe in Africa illustrates the problem of introduction of a sympathy bias into the numbers. Photo courtesy of the United Nations.

Mathematical models can be used to boost causes, both bad and good. A troublesome example of *good cause modeling* is the prediction and monitoring of the spread of HIV/AIDS around the world, especially in Africa, where the disease is taking its worst toll (figure 2.3). UNAIDS, a sub-agency of the United Nations World Health Organization, takes the responsibility for tracking the disease, which it does in large part through the use of mathematical models. UNAIDS now claims that 30 million Africans suffer from the disease. Rian Malan, a descendant of the Malans who instituted apartheid in South Africa, author, anti-apartheid activist, and now an investigative South African reporter, argued in a startling article in the December 14, 2003, *Sunday Telegraph* that the UN models may have distorted the extent of the AIDS epidemic in Africa.

Quantitative mathematical models are universally used to keep track of and to predict the future courses of diseases. But, of course, models require extensive *ground-truthing,* or *field-checking.* In most of southern Africa, record keeping is poor to nonexistent, and with the exception of South Africa there simply is no dependable real-world information to run checks on model results. UNAIDS predicted (in hindsight) that 250,000 South Africans died of AIDS in 1999. This figure was determined by the use of the *Epimodal Model*, the same model that was used to predict AIDS deaths all over Africa. Although the number who died of AIDS is unknown, according to Malan it is accurately established that 375,000 South Africans died of all causes in 1999. The number of AIDS victims predicted by Epimodal is far too large a proportion (two-thirds) of the total deaths. Other public health scientists, using the *ASSA 600 model* (Actuarial Association of South Africa), predicted (again in hindsight) that 143,000 South Africans died of AIDS in 1999. In 2001 the "much advanced" *ASSA 2000* model concluded that there must have been 92,000 AIDS deaths. A run of the new *MRC model* (Medical Research Council) came up with 153,000 deaths in 2001–2002 from AIDS in South Africa.

This is not to negate the importance of AIDS modeling, for the disease is a dreadful global plague that kills before middle age and has left orphans by the many thousands. Indeed, there are real difficulties in determining AIDS death rates because the weakened immune system can result in death from a number of other causes. In addition, doctors may not report the involvement of HIV in order to spare stigmatization of relatives or to prevent invalidation of insurance policies.

Robert Guest, in his book *Shackled Continent* (2004), argues that two kinds of orphans are produced by AIDS—the young and the old. In most African societies the elderly expect their children to care for them in their final years. Instead, the old ones are caring for their dying children and then inheriting their grandchildren. The tragic AIDS-related 1995 death of Nelson Mandela's son brought home to South Africans that no one is safe from the disease. In Durban, South Africa, where the disease is particularly widespread, there are now 600 funerals per weekend compared to 120 five years ago, and graves are being "recycled." And the worst may be yet to come, as the disease appears still to be on the increase in Africa.

But the experience in South Africa suggests that the AIDS disaster may not be as far advanced as previously assumed by the United Nations. Certainly this is a point worth considering, because research on other, more ravaging diseases in Africa, such as malaria, is said to be underfunded because of the anticipated AIDS calamity.

Malaria experts say that 900,000 deaths from malaria occur every year in sub-Saharan Africa. Seventy percent of the dead are children under five years of age. Where did the numbers come from? The estimates of both of these dreadful diseases in Southern Africa suffer from the same lack of local health records. The models have a poor database.

The possibility that a true global human disaster is just around the corner unfortunately provides an unparalleled opportunity for modeling that jacks up the numbers to draw attention and funding. Failure to make a simple reality check allowed the results to become accepted "facts." The apparent sophistication of the models dampened criticism, as did the huge outpouring of sympathy for the afflicted.

In this case the models are probably perfectly good, but the answers they come up with are problematic because:

- the database was poor;
- the models were polluted by a huge "sympathy" bias;
- no one looked back.

Reporter Malan noted: "They told me that AIDS had claimed 250,000 lives in South Africa in 1999 and I kept saying this can't possibly be true. What followed was very ugly—ruined dinner parties, broken friendships, ridicule from those who knew better, bitter fights with my wife. After a year or so she put her foot down. 'Choose,' she said. 'AIDS or me.'" He dropped the subject for more than a year but couldn't resist returning to the question, presumably with his wife's reluctant approval. Malan discovered the going-with-the-grain truth about models, that modeling results are easier to live with if they follow preconceived or politically correct notions.

Sometimes the results from mathematical models are pushed aside by public opinion for reasons bearing no relationship to the veracity of the models. *Epidemiology models* provide standard and widely applied methods to evaluate the causes of human illnesses. *Relative risk* is determined by comparing one population that is affected by something (cigarette smoking, coffee drinking, polluted water consumption, and so on) with the general population. The models must take into account a complex array of *confounders* that could affect the results, such as age, sex, economic status, race, location, allergies, and nationality, among others. Still, the results of these data-rich *statistical models* are widely accepted. Statistical models are based on the assumption that past behavior of a system is a guide to future behavior.

The secondhand smoke (SHS) problem is an example of the public's refusal to accept model results because the results are unpopular.

This is buttressed, of course, by the fact that health problems related to smoking are indisputable and widely recognized and that public tolerance for SHS has rapidly diminished in recent years. Imagine the response of a crowd in an elevator when the big guy in the opposite corner lights up a cigarette!

In 1992 the EPA released its famous report classifying SHS as a class A carcinogen. In this example of *politically correct modeling*, the EPA announced that secondhand smoke was responsible for approximately 3,000 lung cancer deaths each year in nonsmoking adults. On the other hand, in 1998, on the basis of a study involving people from six European countries, the World Health Organization reported no significant cancer risk from SHS.

The EPA report has come under intense criticism. It was based on a *meta-analysis*, a summary of 33 previous investigations, mostly epidemiological mathematical model studies, by others, mostly non-EPA scientists. The number of studies actually used was reduced to 11, yielding a risk factor of 1.19. A risk factor of 3 or 4 is usually required before the EPA considers something a risk to humans. The EPA had announced the 3,000 annual American cancer deaths figure before the study was completed, and when the study did not back up the numbers, it doubled the statistical margin of error to come up with something close to 3,000 in order to save the day. The EPA increased the size of the error bars, thus "enclosing" the number 3,000 between the plus and minus extremes of the prediction.

In 1998 federal judge William Osteen declared the EPA study to be null and void. He noted there was evidence in the record that the EPA had cherry-picked its data (chose only the most favorable studies) and the agency was "publicly committed to a conclusion [3,000 lung cancer deaths from SHS] before the research had begun." In a 2003 speech, Michael Crichton, author of *Jurassic Park* and the highly controversial anti-global-warming novel *State of Fear*, called the EPA study "openly fraudulent science." EPA administrator Carol Browner responded to the judge's 92-page scolding of the agency that "the American people certainly recognize that exposure to SHS brings a whole host of health problems" (probably because of the EPA campaign against SHS). "Consensus trumps science," says Crichton.

SHS opponents routinely claim that the models prove a strong cancer health risk. Not true. Such dishonesty is accepted by our society because the cause (prohibition of SHS) has become a moral issue, not a scientific one. In addition, there are real and significant health problems associated with SHS, including heart disease, pneumonia, and bronchitis, especially among children and asthma sufferers.

Useless Arithmetic on Wall Street

Howard Kurtz and William Sherden, along with many others who have written about stock market prophecy, give innumerable examples of erroneous stock market predictions' being presented with great confidence, the aftermath of which produces no loss of prestige to the failed analyst. Phillip Tetlock in his recent book, *Political Judgment,* shows with statistical analyses that experts in general, and experts on the stock market in particular, are no better than educated non-experts at predicting the future.

Hope springs eternal, however, and market prediction is a field that is becoming ever more quantitative. In 1997 the Nobel Prize for economics was awarded to Myron Scholes and Robert Merton, who, collaborating with economist Fischer Black (who died in 1995), developed a mathematical model for stock market derivatives. Black and Scholes derived the original equation, and Merton is said to have improved it in such a way as to make the model applicable in the real world of Wall Street. The equation involved four variables: duration of an option, prices, interest rates, and market volatility.

The Nobel Prize that year was controversial from the start, although few doubted the genius of the equation. The controversy arose over the question of whether helping rich people get richer was elevating mankind in the sense of Alfred Nobel's original intentions. The next year, possibly in atonement for such insensitivity, the Nobel Prize Committee voted for Professor Amartya Sen, known best for his work on the causes of famine, poverty, and social inequality. The Nobel Prize Committee said Sen "has restored an ethical dimension to the discussion of vital economic problems."

The Black-Scholes equation was widely adopted to calculate the value of options in complex derivative dealings. Derivatives are financial instruments that have absolutely no value on their own, but instead "derive" their value from other assets. The use of derivatives reduces risk and uncertainty in profit, for example, the risk of unexpected price fluctuations. There are derivatives that are contracts or obligations for future delivery, called futures, and there are derivatives that give an opportunity (but not an obligation) to buy or sell at an established price, called options.

The Nobel laureates Scholes and Merton were founding partners in the now infamous Long-Term Capital Management (LTCM) hedge fund that helped fuel the explosion of derivatives trading on Wall Street. At its height, LTCM was the darling of Wall Street, a monetary fund comprising a dream team of Nobel Prize–winning founders and complex financial models who seemingly had developed a clean, highly rational way to earn high returns with little risk, using models and supercomputers to identify investments. Some book titles give clues to the fate of LTCM. The story is told in *Too Big to Fail*, by Kevin Dowd, and in *When Genius Failed*, by Roger Lowenstein. In 1998 the fund that was "too big to fail" suffered catastrophic losses that threatened the stability of money markets worldwide. Lowenstein said the cause was "the disease of perfect belief."

According to Lawrence Summers, former secretary of the U.S. Treasury, "The efficient market hypothesis is the most remarkable error in the history of economic theory." Yet two underlying assumptions behind LCTM's market models were

- that markets are always liquid (e.g., you can always sell an asset at a reasonable price); and
- that markets are efficient and they tend toward equilibrium.

For four years, starting in 1994, LTCM showed incredible returns of about 40 percent per year. Stephen Rhodes (a pseudonym) notes that with about 100 employees, LCTM made more money than McDonald's global hamburger business. All this money and not a useful product in sight.

In the global economy today, international markets are closely linked. A trend in one nation's market can quickly spread to the next. The demise of the LTCM hedge fund began on August 17, 1998. Russia defaulted on its debt, and the worldwide financial markets lost their logical order. Investors fled to more-secure investments, and the firm lost about $3.6 billion in five weeks. On one single day, the firm lost $550 million. The collapse of the hedge fund brought little sympathy from the American media. "We're So Rich, We Can Be Dumb," headlined the *San Francisco Chronicle*.

The collapse of Long-Term Capital Management threatened to create a panic on Wall Street, since many major banks had lent it and other such funds huge sums of money. Almost 50 percent of the world's top banks were involved in rescuing the hedge fund. The consortium gave LTCM $3.6 billion in exchange for 90 percent of the firm. Shareholders retained a 10 percent holding, valued at $400 million, and the dream team kept their jobs. Unfortunately, as some in the media have noted, by sparing shareholders, creditors, and fund managers some of the pain of the loss, we seem likely to see a repeat of the behavior that produced the crisis in the first place.

Economic models applied to the stock market do not work because human emotion and action are unpredictable. Is it not obvious that the stock market is not predictable? It shouldn't surprise us that panic, over-confidence, underconfidence, fraud, ignorance, success, and all kinds of other aspects of human nature control the market.

The lesson to be learned from the Nobel Prize–winning equation and its application by LCTM was forcefully expressed by financial guru and founder of Numa Financial Systems, Ltd., Stephen Eckett, who said, "I regard the Black-Scholes model as one of the most dangerous inventions of the twentieth century. This is not to blame Black and Scholes obviously: the danger is always in the application. But what happened was that one single equation—and mathematically the model is simple—seemed to offer the possibility of quickly understanding and controlling derivatives risk. This encouraged thousands of banks to employ bright mathematicians who had little knowledge of the financial markets but nonetheless started furiously programming their spreadsheets on which billions of dollars were gambled."

Cathy Minehan, president of the Boston Fed, is quoted as saying, while introducing a behavioral economist, "All our models and forecasts say we will have a better second half. But we said that last year. Now don't get me wrong. Mathematical models are wonderful tools. Standard economics would argue that people are better off with more options. But behavioral economics argues that people behave less like mathematical models than like—well, people."

The scandalous bankruptcy at the Enron Corporation holds modeling lessons as well. Economist Keith Cooley describes one of the ways that Enron was able to jack up its apparent profits: "At the heart of the so-called innovative trading at Enron was an accounting rule. When Enron agreed to supply power to a company or municipality at a fixed-price contract, it made projections on the level of future prices and the likely profit over the lifetime of the contract. Under the accounting rule in question, it was then able to report that profit as soon as the contract was signed rather than booking the gains over time."

The problems with Enron's prediction of profits under newly acquired contracts were

- the models were " undisclosed";
- the predictions were always highly optimistic;
- credence was provided through approval by an "independent" accounting firm (the firm, Arthur Anderson, lent the model results an air of credibility, but it had to sell itself off in pieces when the scandal unfolded);
- no one looked back.

A Look Back

Modeling equations are sometimes modified and altered (*tweaked* or *tuned*) until the model correctly "predicts" an already known natural event. Frequently in the modeling literature, however, this is considered a model application or prediction. Although the model may have reproduced something in nature, it is a jimmied equation, one that was adjusted bit by bit to fit a single event or to arrive in the approved range. According to Peter Haff, Duke University model critic, model philosopher, and physicist turned geologist, this approach is better termed *model development*. Haff notes that modeling when the outcome is already known is not the same as a true model prediction before an event occurs. It is in no sense a model prediction.

A good tweaked model example is the modeling of the artificial floods that took place in the Grand Canyon in 1996. Water was purposely released from the Glen Canyon Dam in Colorado to imitate the floods that occurred before the dam smoothed out the peaks in flow volumes (figure 2.4). The problem was that the dam had smoothed out floods and had also trapped almost all the sand coming down the river. In addition, the river was expected to become more difficult and dangerous for rafters because mounds of sediment ranging up to boulders in size, brought to the river by flooding tributaries, were staying in one place, piling up and creating dangerous rapids. Normally, the floods flattened out these rock piles.

The hope for this experiment was that the floods would leave behind new sandbars, just like the old floods once did. Sand would be derived from the stream channel, and the new sandbars would provide much-needed new campsites for river rafters and stream habitats for native fish species that are fast disappearing.

The 1996 experimental flood, probably the first of many to come (a second release was carried out in 2004), didn't completely succeed (politicians declared it a success, but scientists had a different view). Few new sandbars were formed. After the fact, however, geologists were

Figure 2.4 An artificial flood in progress. Water is being released at a high rate from the Glen Canyon Dam in an attempt to provide additional sandbars for river boaters in Grand Canyon. Predictive models of the sandbar configuration that resulted from the water release were unsuccessful. Photo courtesy of the U.S. Geological Survey.

able to tweak the model and come up with the same sandbar configuration on paper that actually remained on the canyon floor after the flood. The modelers confidently suggested that the model would now be useful to predict what will happen in planned future water releases from the dam. Tempting as it might be to believe that the model was now valid, agreement between the model and a single event is not an indication of model validity in a complex system, as noted by Naomi Oreskes. The modelers had developed a new model by tweaking the old one, but it probably won't predict sandbar formation in the next flood. It probably won't be even close.

Today's scientists have substituted mathematics for experiments, and they wander off through equation after equation and eventually build a structure which has no relation to reality.

—*Nikola Tesla, inventor and electrical engineer extraordinaire, 1934*

chapter three

yucca mountain

a million years of certainty

Waste Disposal: A Troubled History

The development and use of nuclear technology began in the early 1940s. Americans grandly entered into a nuclear age that ended a world war and promised permanent supplies of cheap energy. Many cultural images from this time linger with us today: dancing the atomic boogie, drinking atomic cocktails, building backyard bomb shelters, and practicing "duck and cover" drills in schools. Even today the mushroom-shaped cloud remains the high school symbol for the "Bombers" of Richland, Washington, located near the Hanford nuclear plant.

Over time, our perception of the bright promise dimmed as the hazards and by-products of nuclear use became apparent. Some nuclear waste products produce radiation that persists for long periods of time; other waste products can be used to make nuclear weapons. The antinuclear movement has been a vocal, visible presence in the United States for decades. Government failures are largely responsible for the current high level of skepticism that the American public holds for our regulation and management of both the nuclear power industry and the country's defense holdings.

Over the years we have learned that virtually all of the nation's plants that produced or assembled atomic bomb components leaked significant amounts of radioactivity without informing the nearby public. Even the Congress was repeatedly and falsely assured that our nuclear weapon plants were safe and clean, though perhaps the Congress didn't look too closely because the prevailing attitude was that we were in a nuclear arms race for our very survival.

On the evening of December 2, 1949, 11,000 curies of radioactive iodine were intentionally released from the aforementioned Hanford plant. The release, part of Operation Green Run, was one of several in the 1940s. This particular release became public when changes in wind direction, combined with a rare desert rainfall, dumped radioactivity onto the small town of Richland. The operation's backfired results were revealed when workers, entering the plant, set off radiation detectors intended to detect radiation unknowingly encountered by workers exiting the plant. Local radio stations announced that children should stay indoors for a while, that pets should not be allowed to drink from puddles, and that the cause of the problem was a bomb blast at the test site in Nevada.

The radiation was allowed to spread across the state, without warning to those in its path, as a test of American capabilities to detect Soviet Union nuclear activities. The real purpose of the operation was not revealed publicly until 1987, at which time it was also learned that a total of 430,000 curies of iodine was released in more gradual fashion from 1944 to 1947. The radiation was ingested by "downwind" dairy cow herds, producing milk to be consumed by the children of south central Washington.

The March 28, 1979, mini-disaster of Three Mile Island and the truly catastrophic Chernobyl nuclear power plant failure in the Soviet Union on April 26, 1986, dampened any remaining public enthusiasm about nuclear power. Though our cultural celebration of nuclear power may have ended, the production of waste products continues unabated. America's 103 nuclear power plants provide 20 percent of our electricity. And 40 percent of the U.S. Navy's fleet is nuclear-powered. More than 40,000 metric tons of high-level nuclear waste and spent nuclear fuel are temporarily held in 131 locations, throughout thirty-nine of the fifty states. Three-quarters of these locations are within fifty miles of major population centers. More than 161 million people live within seventy-five miles of temporarily stored nuclear waste (figure 3.1). And the waste

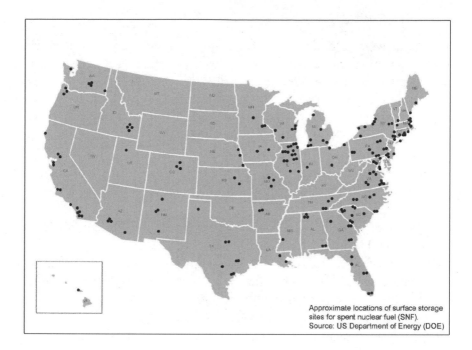

Figure 3.1 Locations of surface storage sites for high-level radioactive waste around the United States. Most of the sites are at operating commercial power reactors. Others include military and research repositories, wastes at plutonium production facilities, and shut-down commercial reactors. Clearly there is a preponderance of "temporary" storage locations east of the Mississippi River. Diagram courtesy of the U.S. Department of Energy.

accumulates by an estimated additional 2,000 tons a year, created from our heavy reliance on nuclear technology for power production, medicine, research, and defense purposes. Assuming that our power plants continue operating under their current licenses, the stockpile of nuclear "garbage" will total 60,000 tons by 2010. Many of the nuclear-waste-generating facilities are reaching their storage capacity and face legal restrictions on construction of any additional storage capability.

The U.S. government is required both by law and by simple necessity to do something with the waste. For many decades there was a naive assumption that when the need arose to permanently dispose of spent fuel and waste, the technology to do so would then be available. As a result, the policy, organization, and infrastructure for waste disposal have failed to grow at the same pace as the nuclear industry.

One example of this is the problem of the waste storage tanks at the Hanford plant. Large steel/concrete waste storage tanks were constructed beginning in the late 1940s and continuing into the 1960s.

The tanks were intended to have a fifteen-year life span, after which it was assumed that a new storage strategy would be in place. Unfortunately, the Atomic Energy Commission (the predecessor to the Department of Energy) made little attempt to seek alternatives to the tank storage approach. As the tanks reached fifty years of age, more than three times their design life, leaks began to occur, and the possibility of explosions within the deteriorated tanks has loomed on the horizon ever since. The nuclear regulators followed an irresponsible path of waste disposal, leaving the problems to be solved by the next generation.

America's inventory of nuclear waste is second only to Russia's. This waste can be hazardous for tens, even hundreds, of thousands of years. Stored radioactive waste could have long-term health hazards that far exceed the known duration of civilization. Not surprisingly, the Office of Homeland Security believes that the storage sites could be prime targets for terrorists and therefore represent a significant hazard to millions of citizens. The storage of nuclear waste is a huge political and physical problem, as well as a possible national security issue.

Until the 1970s, the recycling of spent nuclear fuel was also considered a potential answer to the waste storage problem. Reprocessing spent nuclear fuel can recover unreacted uranium, which is then available to be used again as nuclear fuel. The remaining waste can be solidified before disposal. Unfortunately, plutonium, which can be used to make bombs, is also isolated by this process. During the Ford and Carter administrations, a nuclear nonproliferation policy was established and recycling was prohibited, in order to avoid the production of plutonium. It was hoped that the American approach would set an example for the world to follow.

In 1974 our proliferation fears were proved real when India tested a nuclear weapon it had developed using plutonium from reprocessed spent fuel. The development of the India bombs, followed by Pakistan's and perhaps North Korea's and Iran's, can be construed as a failure of our nonproliferation policy. Abdul Qadeer Khan, Pakistan's father of the bomb, didn't help matters by selling how-to-make-the-bomb secrets to Libya, North Korea, and Iran.

Would there be a high-level nuclear waste problem if this nonproliferation policy were reversed? Probably not, or at least it would be a much smaller one. The French and the British use the recycling option, and they do not have the critical storage issues that we have.

The Long Search

Disposal in space, burial under the ocean floor in abyssal plains or in polar ice, and deep well injection have all been other theoretical storage options considered in the past. Most nations today favor the concept of a *geological repository*, burial of radioactive waste in underground facilities within national boundaries. A *storage site* implies a temporary arrangement, but a repository is permanent. Scientists and policymakers believe that geological disposal limits the need for ongoing control and cost to future generations.

In the United States, underground salt beds were long considered the ideal location for permanent storage of nuclear waste. Most salt deposits are found in relatively stable areas with a low potential for earthquakes. Salt is preserved in rocks where water is absent, or the salt would have long since dissolved. A salt deposit in southwestern New Mexico is currently a geological repository for low-level, defense-generated waste, but the state successfully fought plans to store high-level waste at the same location. A salt mine in Kansas was once studied as a permanent storage site but was also eventually rejected. In both cases, it seems that politics (the relative power of the Nevada, Kansas, and New Mexico congressional delegations) determined the rejection of the salt sites.

Belgium is considering a deep clay formation as a candidate for a geological repository. China's potential site lies in the Gobi Desert, in the granite beneath the water table. Russia has identified two potential sites and has eventual plans for a total of four geological repositories. Canada, France, Japan, and the United Kingdom have not formally selected a candidate for a repository.

In the United States, the search for a permanent location for both commercial and defense waste was formalized in 1982. The federal government has been collecting a tariff from waste-generating utilities in exchange for a promise to take the waste. The Nuclear Waste Policy Act (NWPA) called for an exhaustive study to screen potential permanent storage sites and set a deadline of January 1998 for the federal government to begin accepting waste for safe storage. The NWPA also called for two separate waste repositories—one to the west and one to the east of the Mississippi River—to assure regional equity within the nation. The idea for a second repository ended, however, after many NIMBY (not in my backyard) objections from the politically powerful eastern states. Among other roadblocks to an eastern repository was a law passed by Congress outlawing storage in granitic rocks, which, of course, happened to be the rock type being considered east of the Mississippi River.

To make sure that no single agency would exert full control over such an important facility, Congress established a system of checks and balances: The Environmental Protection Agency (EPA) would set health and safety standards, the Nuclear Regulatory Commission (NRC) would evaluate whether those standards had been met and issue a construction license for repositories, and the Department of Energy (DOE) would build and manage the site. Meanwhile, the DOE's science became so politicized, and of such questionable quality in several other regards, that Congress established the Nuclear Waste Technical Review Board to oversee the agencies' research efforts.

The Yucca Mountain geologic repository site "won" after a long and stormy search that considered and discarded many other sites for both political and scientific reasons. Sites in three western states were selected for a more intensive study in 1986. These sites were Yucca Mountain, near Las Vegas, Nevada; a basalt formation under the Hanford Nuclear Reservation in Washington State; and the salt deposits of Deaf Smith County, Texas. The NWPA required that an environmental assessment be completed to make the final site selection. Heavy national politics ensued, and in 1987 an amendment to the NWPA left Nevada as the only remaining choice.

The mountain is about a hundred miles northwest of Las Vegas, adjacent to the federal Nevada Test Site, where the government has conducted more than eight hundred nuclear bomb tests. It is a flat-topped volcanic ridge, running about six miles north to south, that from the air appears to slither across the desert like a snake (figure 3.2). A five-mile-long tunnel has been bored into the mountain in anticipation of ultimate legal approval of the project. The site has now been studied, poked, and prodded to the tune of $4 billion.

The EPA set a stringent requirement for the permanent disposal site: that leakage can expose humans to no more than 15 millirems of radiation in any given year for the next 10,000 years. If it wasn't obvious at the time the regulation was written, it should have been clear by the year 2004 that the EPA's demand for 10,000 years of certainty—based, of course, on mathematical models—is both absurd and unattainable. The year 2004 was the year that a federal appeals court ruled that instead of 10,000 years of certainty, the successful functioning of the repository must be assured for hundreds of thousands of years, up to a million years! The decision

Figure 3.2 Aerial view of Yucca Mountain, Nevada. The proposed repository site is in a remote arid region where the groundwater table (the level of permanent water saturation) is 2,000 feet below the surface. Yucca Mountain is in a closed basin, where water does not flow to a river. Photo courtesy of the U.S. Department of Energy.

was based on the fact that peak dosage or the maximum radiation from much of the waste would occur well beyond 10,000 years from now. But increasing the requirement to predict the fate of the waste for 100,000 or more years is simply bizarre. The court may as well have required that all the nuclear material be buried on Venus.

The new Yucca Mountain evaluation now encompasses a time span ten or more times that of the presence of humans in the Americas, and longer than Homo sapiens has been on the earth. It will likely encompass at least two major advances and retreats of glaciers, with the accompanying huge changes in climate. Is the legal system completely disconnected from reality?

The State of Nevada is convinced that the site location process was largely political. Locally the legislation is known as the "Screw Nevada" bill. Nevada opposes the plan for permanent storage and will be fighting the project in court for years to come. Meanwhile, as this process has played out over decades, commercial nuclear utilities have paid more than $15 billion into a nuclear waste fund. The DOE is already years past the deadline to begin accepting wastes. Frustrations and lawsuits have mounted over decades of no progress.

The Waste

Wastes are classified as low level, middle level, or high level, according to the amount and types of radioactivity in them. *High-level waste* may be the spent fuel itself or the principal waste from reprocessing the fuel. While high-level radioactive waste is only 3 percent of the volume of all radioactive waste, it holds 95 percent of the radioactivity. It contains the highly radioactive fission products and some heavy elements with long-lived radioactivity. It generates a considerable amount of heat and requires cooling, as well as special shielding during handling and transport. This is the stuff of Yucca Mountain.

Middle-level and low-level wastes are usually handled within the borders of the state where they have been produced. *Middle-level waste* includes that from reprocessing of fuel rods and from the decommissioning of power plants. *Low-level waste*, generated by hospitals, research labs, and power plants, includes such things as lab coats, gloves, contaminated tools, piping, and filters from air-purification units

The largest category of waste to be stored in the repository is spent fuel from power plants. Nuclear power reactors use fuel made of ceramic pellets (the size of a little finger) of enriched uranium sealed in tubes. The tubes are bundled together to form the nuclear fuel assembly. One pellet has an amount of energy equivalent to one ton of coal. Heat from the tubes forms steam, which turns giant turbines. After the fuel is spent, it is removed from the reactor and placed in pools of water contained in steel-lined concrete basins. The spent fuel eventually may be moved to steel or concrete dry storage containers or bunkers. Forty years after removal from the power reactor less than one-thousandth of the initial radioactivity remains, and it is much easier to handle.

Other categories of waste include perhaps 3,000 metric tons of spent fuel from reactors on naval ships, research reactors, and waste from weapons production. Liquid waste derived from rod manufacture and weapons production amounts to about 30,000 tons. Plans are to turn this material into glass (vitrify it) before storage. Another small but important category is excess plutonium from weapon production.

The Site

Yucca Mountain has many appealing features as a permanent storage site. Located on a large federal land reservation, it is far from a major population center (figure 3.3). Only eight people live within fifteen miles, although the future may change that. The Yucca Mountain environment is arid, with an average of about six inches of rainfall a year. Nearly all of the precipitation evaporates. The surface water drainage system is a closed hydrologic basin, so surface water does not flow from Yucca Mountain into major river systems. There are no perennial streams nearby. And, most important, the water table is incredibly deep.

The State of Nevada believes Yucca Mountain should be disqualified as a potential site because of three factors: the potential for human intrusion, the potential for faulting and volcanism, and the rapid groundwater travel time from the repository to the environment.

Water in the repository is the primary medium by which the radioactivity could escape the "system." First, it can accelerate the corrosion of the waste-storage packages. Second, as water contacts the waste, radionuclides could become mobile and be released into the nearby ground and surface waters. Water is anathema to the storage strategy of Yucca Mountain and must be kept away from the wastes as long as possible.

The mountain is composed of alternating layers of welded and nonwelded tuff, a rock formed from volcanic ash estimated to be between 11 and 13 million years old. The degree of welding determines how water moves in the various units: through fractures in welded units and through the matrix in nonwelded units.

The repository is being built in a horizon about 1,000 feet below the ground surface and approximately 1,000 feet above the water table, in the thick unsaturated zone within the mountain. The unsaturated zone is thought to provide a barrier against potential rises in the water table because of unforeseen natural events. If everything went wrong and the waste was introduced in large quantities into the groundwater, it would flow toward Amagosa Springs, to the south of Death Valley, currently a very lightly developed area. Under no circumstances could ground or surface waters contaminated by the failure of the site flow to Las Vegas.

The project is to dig 100 miles of tunnel, called emplacement drifts (figure 3.4), and fill them with thousands of waste storage containers, each about the size of a tank on a tanker truck. Remotely operated trains and robots would do the dirty work. The project could eventually hold 77,000 metric tons of waste. A thick titanium drip shield would top

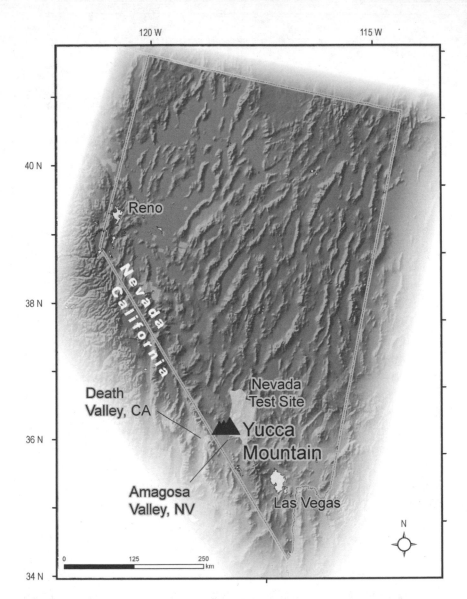

Figure 3.3 Yucca Mountain would seem to be an ideal remote location. It is close to the already contaminated atomic bomb test sites, and if the repository fails to behave as predicted, contaminated wastes will flow toward Amagosa Springs and Death Valley, away from populated areas. Figure by David Lewis.

waste canisters, which would be made of stainless steel and other durable metals. The shields, an afterthought to correct for modeling inaccuracies, would protect against both water seepage and falling rocks. After at least fifty years (and likely more), the repository would close forever, the entrances would be blocked, and the site would be marked to warn future generations of the danger lurking below.

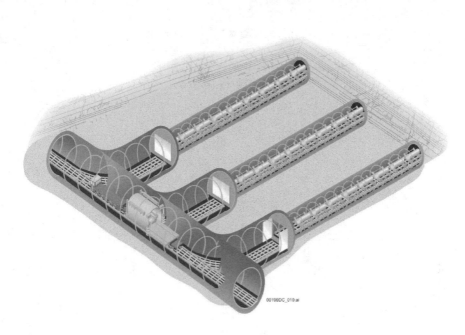

Figure 3.4 Radioactive waste at Yucca Mountain will be stored in underground emplacement drifts, three of which are illustrated in this diagram. Diagram courtesy of the U.S. Department of Energy.

The Models

The DOE is using a *total system performance assessment* (TSPA) methodology to evaluate how the Yucca Mountain Project complies with EPA standards. The models seek to predict the behavior of both the natural system (the mountain) and the engineered system (the casks and the preventative drip shields). In some of the literature for public consumption the TSPA is visualized as a pyramid, with the base containing data and observations of the natural and engineering portions of the system. Models based on the data make up the next layers of the pyramid (figure 3.5). The overarching model integrates many separate modeling efforts, each differing in complexity and scale and feeding into a more comprehensive model higher up on the pyramid. There are 13 comprehensive mega models or model clusters based on 286 individual models. Clusters include models predicting future climate, infiltration, percolation, and water behavior in drifts. There are thousands of input parameters, hundreds of thousands of lines of equations in hundreds of computer codes, and hundreds of linked mathematical models in the system: complexity built upon complexity, assumption built upon assumption.

Predicting the Future of Yucca Mountain

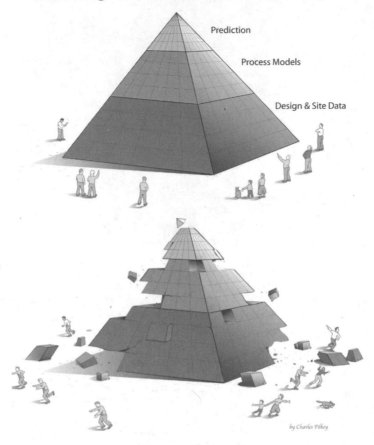

Prediction

Process Models

Design & Site Data

by Charles Pilkey

Figure 3.5 The Department of Energy views the modeling effort at Yucca Mountain as a pyramid. At the bottom are field observations. In the second layer are the hundreds of mathematical models that predict how natural processes will work over very long periods of time. At the top are the models that put it all together to predict the behavior of the repository over a long period of time. But a pyramid founded on limited data and faulty models projecting far into the future can never survive! Drawing by Charles Pilkey.

The TSPA estimates the likely behavior of the repository using what are said to be conservative assumptions to compensate for uncertainty. It also attempts to quantify with mathematical models the probability of natural events (such as earthquakes and volcanism) that could weaken the waste canisters. The current methodology uses predictive science almost to the exclusion of other approaches.

Norm Christensen, longtime member of the Nuclear Waste Technical Review Board and former dean of the Nicholas School of the Environment at Duke University, is highly skeptical of the predictive capabilities of the models used by the DOE. In Christensen's view, the models have been useful to structure our understanding of the problem and to organize our thinking and our approaches to understanding the future of stored waste. The models can provide information on the relative importance of parameters, point out gaps in understanding, and provide order-of-magnitude answers for many of the problems facing the repository in the long-term future. But the models, which are mostly deterministic, cannot provide the accurate prediction standards required by the TSPA. They can't do what the scientists long ago promised agency officials could be done.

As is the case with all models of natural earth surface processes, two broad categories of uncertainties are inherent in the Yucca Mountain modeling. One has to do with chance and refers to the variability of the many factors that make up a physical process. This is the *ordering complexity* factor. We argue that quantitative modeling of complex systems is impossible. Even if one could characterize accurately all of the parameters that go into a process, one never knows the order of their occurrence, their magnitude, or their duration.

The second category of uncertainty has to do with the state of our knowledge. Complexity cannot be avoided or sidestepped in any way, but the level of knowledge about processes can usually be improved with field and laboratory studies. The questions posed below are a small sampling of the uncertainties on the modeling table at Yucca Mountain.

• As radioactive waste begins to flow into the surrounding rock, how far and fast will it go? To determine flow rates, the porosity and permeability and slope of the various rock units must be known.
• How do porosity and permeability vary laterally? Flow rates can vary within three and sometimes more orders of magnitude in single rock types.
• Where are the faults and fractures that could either speed up or slow down flow rates?
• What will the shape of the dispersed body of waste be? How much lateral flow will occur?
• What if the climate changes (a virtual certainty)?
• What will increased rainfall do to groundwater flow rates?
• What if the groundwater table moves up and water encompasses the waste tunnels?
• What reactions will occur in the unsaturated zone (above the groundwater table)?

- What reactions will occur in the saturated zone (below the ground-water table)?
- How fast will these new chemical compounds move in the ground-water?
- What is the solubility of the new compounds in the groundwater?
- What will chemical reactions between the rock and the waste do to the rock's permeability and its capacity for groundwater movement?
- What temperature will the escaping waste be, and what will that temperature do to flow rates? to chemical reactions? to solubilities of the various compounds?
- How fast will the protective titanium "roof" deteriorate?
- How soon will the canisters containing the wastes begin to leak?

There seems to be an unstated assumption in the current approach to modeling the repository that technology and knowledge will not advance in the future. Thus, we must design a solution for 100,000 to 1,000,000 years today. Would it not be simpler and more logical to design a 100-year solution? In those 100 years, scientists and engineers would continue to look for ways to store the material for a longer period. Assuming they succeed, and it seems reasonable that they will, a new method of storage would then be used, which might be good for, say, 250 years. And so it would go until a safe, permanent solution is discovered. Of course, this approach leaves the management of the waste to future generations, with no guarantees.

The Saga of the Percolation Flux

The downward movement of water, called infiltration, is controlled largely by gravity. *Net infiltration* is water that penetrates to sufficient depth that it is not removed from the ground by either evaporation or transpiration by plants. The rate of downward movement of water in the unsaturated zone above the water table is called *percolation flux*. This is one of the key parameters affecting the viability of the Yucca Mountain performance. It is expressed as millimeters per year. The higher the flux, the greater the likelihood that water will seep into areas where the waste material is stored.

Determining percolation flux is difficult because it cannot be measured directly, especially in an arid area such as Yucca Mountain. It can be estimated through the use of models or indirectly calculated through physical measurements on rock samples. Initial estimates of the rate of percolation flux were derived using data from previous studies of the adjacent bomb test site, on-site rock corings and computer simulations. In various studies between 1983 and 1986, the estimated percolation flux was lowered continuously from 4 millimeters per year, to 1 millimeter per year, and then to 0.5 millimeter per year. Percolation flux rates of 0.02 to 1 millimeter per year became the foundation of the performance assessment mathematical modeling. One of the fundamental assumptions behind such low downward-flow rates was that groundwater was not moving through fractures and faults but rather was moving through the tiny interstices between grains within the matrix of the rock. On the face of it, it was a puzzling assumption.

A change in the flux rate can have enormous impact, because the increase or decrease is nonlinear. In other words, an increase in the percolation flux will produce a proportionately larger increase in the amount of water seeping into the repository.

Beginning in 1996, however, the construction of the tunnel into the mountain offered the opportunity for systematic physical sampling of water in the unsaturated zone. The discoveries from this sampling took project managers by complete surprise and completely contradicted the models.

In late April 1996 the DOE released a report by Los Alamos National Laboratory researchers that documented elevated levels of radioactive chlorine-36 in water within the vicinity of several of the faults uncovered by the tunnel boring machine within the proposed repository, although chlorine-36 is a global marker created by the bomb tests on Bikini Atoll in the South Pacific before 1963. To get 600 or more feet below the surface, where the isotope in the groundwater was discovered in less than fifty years from the time of its production, the radioactive isotope had to have been carried there by water flowing rapidly downward from the ground surface—prima facie evidence that fast groundwater pathways exist at Yucca Mountain (at least in the vicinity of faults). Instead of less than 0.5 millimeter per year, the chlorine-36 derived rate was on the order of 3,000 millimeters per year!

This was, however, a localized observation around fault outcrops and does not mean that the entire water content of the overlying rocks moves down at the faster rate. But, the discovery of these "markers" at and below the level of the proposed repository cast doubt upon the DOE's entire hydrologic model for Yucca Mountain, which had assumed that water traveled extraordinarily slowly through the subsurface, principally by moving through the pores or matrix of the rocks. Such movement was postulated to occur at just a few feet per 1,000 years.

In 1997 the views of seven experts were solicited, and an aggregate distribution of opinions about percolation flux was produced. The expert panel's opinions ranged from 1 to 30 millimeters per year, with a mean (average opinion) flux of 10 millimeters per year. This became the official percolation flux, a product of science by majority vote!

The significance of the chlorine-36 finding is that the DOE's own repository siting guidelines, and the NRC licensing regulations, require a site to be disqualified if it is shown that groundwater travel time through the repository to the accessible environment is shorter than 1,000 years. At Yucca Mountain, this mark was missed by two orders of magnitude! Water from the surface arrived in the tunnel in 50 years, after having traveled through almost 1,000 feet of rock. Some Yucca Mountain scientists are already downplaying this finding, saying that they have discovered only six places where there is fast-moving water and that most rainfall will take tens of thousands of years to seep down that far, which is probably true. Nevertheless, the role of the engineered barriers in the projected long-term success of the project suddenly became more significant. Originally Yucca Mountain was chosen because of its natural characteristics. The plan to safely store waste now depends on engineered barriers, including the titanium drip shields, which were an add-on to the design.

The question remains, how could a number such as percolation flux, upon which dozens, perhaps hundreds, of mathematical models depended, be so wrong for a decade? Directly or indirectly, the number remained as the foundation of a multibillion-dollar modeling effort even though it was based on the questionable, even absurd, assumption that water was not flowing down fractures and faults within the rock. Daniel Metlay, staff member of the Nuclear Waste Technical Review Board, argues that organizational culture was at work: "When faced with the need to resolve uncertainty about percolation flux, the scientists had little organizational incentive to settle on a higher value or, more important, to question whether a lower value was correct."

Where to Next?

We are close to capacity at our temporary storage sites. All our eggs are in a basket that won't provide the needed immediate relief. Spent fuel must lie underwater, on-site at the power plants, for five years before it is cool enough to move safely. Moreover, the repository itself is far from complete, with processing centers still to be built and miles of tunnels to be bored. Once the project is completed, even an ambitious program to ship waste to Yucca will take three or four decades, according to the Department of Energy. An estimated 93,000 shipments by truck and rail will be needed to move radioactive waste sealed in accident- and terrorist-proof containers.

If the DOE's supervision of contractors digging the initial six-mile tunnel is any indication of the agency's organizational competence and concern for safety, perhaps the worries about waste transport are well founded. For the first mile and a half of tunnel, the miners operating the tunnel digging machine were unprotected from dust, and some of them are already showing signs of lung problems, including silicosis. Since silicosis is considered to be completely preventable (with high-quality face masks), it is an astounding mistake in oversight from an agency already under fire for both poor management and weak science. Ironically, the dust could have been somewhat reduced by large-volume water sprays, but scientists were concerned that the introduced water would compromise their studies of the percolation flux.

Individual states are gearing up to oppose the transportation of waste by truck or rail, and the State of Nevada hopes to take advantage of this growing opposition. If Yucca Mountain opens as scheduled, the waste from the temporary storage sites around the nation will begin to roll down our highways and railways in a decade. There are estimates that one in seven Americans lives within one mile of a proposed transit route.

On February 11, 2002, Secretary of Energy Spencer Abraham certified to President Bush that it was safe to build a nuclear waste repository at the Yucca Mountain site in Nevada. Four days later President Bush accepted the certification, stating that an underground facility is a safer alternative for the millions of Americans who now live within seventy-five miles of a temporary aboveground waste facility, that safe storage of nuclear waste is necessary to ensure that the navy's nuclear-powered fleet can operate long into the future, and that safe storage of nuclear waste will enhance public confidence in nuclear power generally, thereby creating more alternative sources of energy for a nation seeking greater energy independence. The DOE is now preparing an application to obtain an NRC license in order to proceed with construction of the repository. The current time line calls for completion of the Yucca Mountain facility in 2010.

More than five decades and $4 billion have been spent to find a suitable site for radioactive waste. The pressure is on, and policymakers may choose to overlook the troubling signs that Yucca Mountain is not a suitable location, at least not according to the government standards set long ago. The standards, however, are being lowered with each troublesome finding at the Yucca Mountain site. The Nuclear Regulatory Commission noted 293 flaws in the Yucca Mountain studies and designs by the DOE.

But, with the benefit of perfect hindsight, we now realize that the repository standards were set impossibly high and expectations from predictive mathematical models were completely unrealistic.

The fact that the current approach to Yucca Mountain feasibility studies is almost exclusively through mathematical modeling must have resulted from some group of scientists and engineers, somewhere, a few decades back, assuring officialdom that it was feasible to predict the fate of stored waste tens or hundreds of thousands of years into the future. And that became the standard to be achieved.

The strict adherence to a prediction timetable allows no flexibility, no opportunity for change, no place for evolution of future plans. The standard is clearly impossible. The modeling approach has no possibility of providing the answer. The models will fail for a number of reasons, among them:

- an impossible predictive time span (hundreds of thousands of years);
- no experience basis for understanding the role of time in chemical reactions of waste or in the degradation of waste containers and the titanium shield;
- climate change, a critical factor and a complete unknown;
- the magnification of errors created by the interdependence of hundreds of models;
- ordering complexity of the natural processes that are involved.

Danish physicist Per Bak, the coauthor of the books *How Nature Works* and *Why Nature Is Complex*, notes that we are constrained by our imagination and can go only to the adjacent possible. Which means that when it comes to prediction, we are locked into our history—we can't stray far from it. We can only project the known into the future. Michael Crichton, controversial model critic of *Jurassic Park* fame, holds a similar philosophy. He notes that a primary future concern of city officials at the beginning of the twentieth century must have been what to do with the increasing amount of horse manure produced by the burgeoning numbers of horses in cities. Who could have foreseen the role in society of the automobile, nuclear power, or the airplane? The application of this principle to Yucca Mountain is obvious.

Per Bak suggests a strategy for coping with the complexities that we encounter:

- Don't predict.
- Adapt.

This is simple and profound advice.

Is There an Alternative?

In contrast to Per Bak's suggested approach, the guiding philosophy at Yucca Mountain has been to optimize and predict. The DOE approach relies on technological solutions and predictive certainty. *Adaptive staging*, which relies on a qualitative scientific understanding of the system, is another solution. This is the approach being used by Belgium in the design of that nation's nuclear repository.

Adaptive staging can be compared to driving a car to work. The driver knows that she is going to drive from point A to point B, and the route is clear. But along the way she is required to make a number of adjustments that are unknowns when she starts out. For example, the driver may have to alter the car's trajectory to avoid potholes, miss pedestrians and road construction equipment, and steer clear of the unexpected car darting out from a driveway.

Use of the mathematical model MARXAN affords a comparison of modeling and adaptive staging. Two Australian scientists, doctoral student Ian Ball and his advisor, Hugh Possingham, devised MARXAN, intending it as a method of determining the best way for governments, foundations, or environmental groups to maximize biological diversity in a region by purchasing land or placing restrictions on land use. One of the more spectacular accomplishments guided by the model was the halting of fishing over a large part of the Great Barrier Reef by the Australian government. In the decades before the model became available, a form of adaptive staging was used. Property was purchased or protected on a year-by-year basis if it sheltered important habitat or species. Property was obtained as it became available, whether or not it was part of a plan. MARXAN, on the other hand, presented an optimal and basically inflexible plan to be carried out for a decade or more. The model's approach to preserving biodiversity, once the data were obtained and the model run, provided a relatively uncomplicated and comfortable solution for those who liked to avoid surprises and complications. It also provided a firm plan that should impress would-be financial contributors.

In 2004 Possingham and two American colleagues evaluated the success of MARXAN by hindcasting the Nature Conservancy's effort to preserve biodiversity on the Columbia River Plateau. According to a September 21, 2004, *New York Times* report, the simple, flexible adaptive-staging approach proved more successful than MARXAN. The advantage of the simple, flexible, and qualitative approach was that it evolved as changes occurred in the area in question, such as greater development density, degradation of property, and ownership changes. The inflexible mathematical model became outdated within a year.

In the application of adaptive staging in the construction of a nuclear waste repository, siting and construction could proceed in stepwise phases. Decisions on how and where to proceed would be made by evaluations at the end of each stage. The focus would be on exploration rather than prediction. Adaptive staging would avoid complete specification of technical requirements at the beginning—the way the project operates now—and would allow the design process to evolve. Such evolution would permit the inclusion of new technology, new ideas as they come forward.

This is the path that Sweden is taking. It is a country that has no dry desert areas like Nevada, and it is a fair assumption that within the next few thousand or tens of thousands of years the country will be covered by glaciers again. The most stable rock type in Sweden is granite, which is everywhere completely saturated with water. A repository site was chosen sixty miles south of Stockholm, where 1,000-foot shafts were constructed. The water at 1,000 feet proved to be completely free of oxygen, so copper, which is stable under such anoxic conditions, was chosen to encase the waste. Eventually it was discovered that a huge concentration of copper-corroding microorganisms exists in the deep water, and that discovery necessitated an additional shielding layer of inert bentonite clay around the copper containers.

Step by step, the Swedes will design their repository over the next few decades, using data from science and engineering and independent of electoral cycles. If only the United States could be so lucky.

Examples of adjustments that could be made if a stepwise approach were applied to Yucca Mountain are:

- Purchase all private property that is in the trajectory of groundwater flow from the worst-case scenario of waste release at Yucca Mountain.
- Design super-safe transportation procedures for waste, such as brief temporary closure of interstates and railroads to other traffic.
- Reduce transportation hazards by providing additional waste repositories on the East Coast or in the Midwest.
- Reprocess some of the fuel rods to reduce waste volume.
- Don't seal Yucca Mountain. Keep it open and alter storage techniques as technology advances.

The rigid design approach employing mathematical modeling of natural processes that is currently being used at Yucca Mountain probably is cheaper and faster. Adaptive staging, on the other hand, allows more input from all stakeholders, early diagnosis of the inevitable problems, and building of trust among all the parties involved (although this would seem to be an impossible dream for the highly contentious radioactive waste debate). Rigidity versus adaptiveness. Per Bak would be pleased with the Belgians and unhappy with the Americans.

Reliance on a pyramid of hundreds of quantitative models leaves the Yucca Mountain project highly vulnerable to criticism. Now the State of Nevada has an unlimited supply of poor assumptions, weak field data, unknown parameters, and poor models feeding into other poor models on which to base its legal challenges. This must be an attorney's dream. If the project is required to go ahead on its scientific merit, the end of the Yucca Mountain saga is in plain sight. The more likely eventuality, of course, given the history of Yucca Mountain, is that politics alone will decide the fate of the repository, and the model results will be a fig leaf providing cover for the politicians.

All the rivers run into the sea; yet the sea is not full.

—*Ecclesiastes 1:7*

chapter four

how fast the rising sea?

Senator James Inhofe, the Oklahoma Republican and University of Tulsa graduate, has been chairperson of the Senate Environment and Public Works Committee, which oversees legislation that could address the problem of excess production of greenhouse gasses. Our nation spends more than $2 billion a year for global change studies, and he is the most powerful American, short of the president, with regard to the initiation and funding of legislation that would form our response to global warming and climate change, including reduction of carbon dioxide emission. The *New Yorker* magazine attributes the following 2003 statement to the senator: "With all of the hysteria, with all of the fear, all the phony science, could it be that man-made global warming is the greatest hoax ever perpetrated on the American people? It sure sounds like it."

This profoundly unenlightened statement says much about the influence of science in our society and the credibility of global change science.

Changing Levels of the Sea: A Perpetual Event

Sometime around four billion years ago, Earth cooled and allowed water to organize and amass instead of boiling away. Some of the water was "juvenile," brand-new water molecules formed from the "outgassing" of hydrogen and oxygen atoms during violent volcanic eruptions. By one billion years ago, the volume of water that makes up our oceans had largely been accumulated.

The basins that hold the ocean's water formed at the same time, and more or less kept pace with the accumulation of water. Earth's lighter granitic materials and heavier basaltic lava-like materials segregated into continents and ocean basins, respectively. The light material formed continents that float like icebergs above the heavier material of the ocean basins. At the same time that the continents were being assembled, they began to drift about. Ocean basins expanded and contracted, and the level of the sea relative to the continents went up and down like a slow-motion yo-yo.

Since the beginning of time on Earth, the level of the sea has been in a constant state of change. It is fair to say that some of the mechanisms of change are well understood, while others are only poorly grasped. Undoubtedly there are some mechanisms of sea-level change that aren't even recognized yet. The dynamics that cause these changes are many—some of them major and global in scale, some minor and local, some carried out over millions of years, and some over minutes and seconds. All sea-level changes are either *eustatic* or *isostatic*. *Eustacy* refers to sea-level changes that happen because of variations in the water volume. *Isostacy* refers to changes in sea level that result from the moving up and down of the earth's surface. Such changes could occur because of local mountain-building forces, subsidence caused by the weight of glaciers, or the compaction of sediments on deltas.

Numerous processes and events drive eustacy and isostacy. The largest long-term changes of sea level occur on a time scale of 100 million years or so, as a result of variations in the volume of the ocean basins (isostatic changes), such as the rise and fall of the giant mid-ocean ridges or the upheaval and sinking of the world's major tectonic plates. When the volume of the ocean basin gets smaller, global sea level rises and the water sloshes onto the continents. Conversely, if the ocean basin volume increases, sea level drops and ocean waters withdraw from the continental land surface.

On another scale, sea-level variations (eustatic changes) can occur on a global scale in a time frame ranging from 100,000 years to the life span of a person, caused by the capture or release of water by expanding or contracting glaciers. The vast continental glaciers occurred during distinct time intervals called ice ages. An ice age is really a time that alternates between cold periods of peak glaciation and warmer periods, like today, when the world is mostly free of ice (except for Greenland and Antarctica). Today, we are a couple of million years into the middle of an ice age known as the Pleistocene epoch. Large tongues and sheets of ice pushed down from the high latitudes in at least nine major and many more small separate episodes, covering at their peak as much as 30 percent of the area of the continents. The miles-thick glaciers accumulated snowflake by snowflake, capturing huge volumes of water. During times of peak glaciation, the exchange of water caused the sea level to drop by as much as 350 feet or more (eustatic change).

The fundamental causes of the ice ages and their huge sea-level changes are various Earth orbital changes, such as the tilt of the axis of spin and the eccentricity of the orbit around the sun. These changes are responsible for changes in the location and intensity of solar radiation on the surface, which in turn determines global climates.

Isostatic changes in sea level can be very important locally, especially in the high latitudes where the land was once covered by glaciers. These are the vertical land movements induced by the load of ice and water on the continents. When massive glaciers flow to the sea, their weight depresses the margins of continents by hundreds of feet. This causes huge but local rise in sea level, as happened along the New England and Maritime Provinces coasts (figure 4.1). When glaciers decay and retreat, the land rebounds and the sea falls back. A present-day example of this phenomenon is Juneau, Alaska, where, because of the removal of the weight of the retreating Mendenhall Glacier, the land is rising and the sea level is dropping.

Sea-level fluctuations related to the advance or retreat of glaciers cause alternate flooding and draining of the continental shelves. When the water level is up, as it is today, the water's weight slightly depresses the edge of the continents. This is one of a number of mechanisms responsible for today's sea-level rise on both the East Coast and the Gulf Coast of the United States.

Lots of uncertainties remain. Until very recently it was thought that the warming of the oceans was probably a more important cause of sea-level rise than the melting of glaciers and polar ice caps or crustal

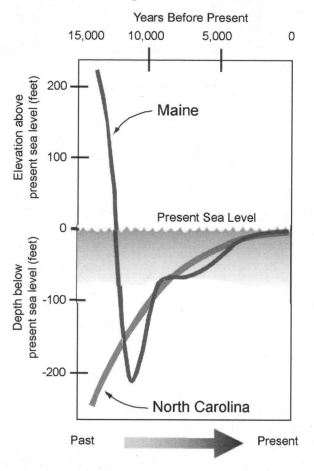

Sea Level Changes in Maine and North Carolina

Figure 4.1 A generalized comparison of the sea-level history of northern Maine and North Carolina over the last 15,000 years. The history of sea-level change in North Carolina for the most part simply reflects the change in ocean water volume resulting from the melting of the glaciers. In Maine the land was depressed (and the sea level was high) because of the weight of the glaciers. As the ice retreated from Maine, the land rebounded, complicating the changes in the level of the sea. Diagram based on information from Joe Kelley.

depression. As water warms up, it expands ever so slightly. In a body of water the size of the ocean, even a slight expansion can cause a significant sea-level rise known as the *steric effect*. If global temperatures of the atmosphere were to stabilize right now, expansion-related sea-level rise caused by past conditions would continue for many years. However, in a paper published in 2004, Walter Munk, a physical oceanographer at the Scripps Institution of Oceanography, argued that there was a mismatch between observed sea-level rise and that calculated from thermal expansion of seawater induced by global warming. He suggested that the steric effect is less important and that the melting of glacial ice is more important than usually assumed by modelers of sea-level rise.

Active mountain-building forces on the steep, mountainous Pacific side of North and South America greatly confuse the sea-level picture. These forces can be violent, often manifested by earthquake and volcanic activity. The western margins of these continents sometimes move up and sometimes move down. Relative sea level in San Francisco and Los Angeles is rising at a rate of a few inches per century, but it is dropping slowly in Crescent City, California, in Astoria, Oregon, and in Sitka, Alaska.

In 1964 the level of the sea along the entire western margin of Alaska's Kenai Peninsula rose two to three feet in an instant, as the land subsided during the Good Friday Earthquake of 1964 (figure 4.2). Along the tropical Pacific Coast of Colombia, earthquakes (accompanied by tsunamis) cause the shorelines to sink three to four feet at a time, resulting in some astounding local sea-level-rise rates of as much as ten feet per century. After the Great Tumaco Earthquake of 1972, the level of the barrier islands fronting the Colombia Pacific shoreline dropped three feet or more along fifty miles of shoreline, virtually overnight. High shoreline erosion rates immediately followed, and small villages were abandoned, their residents forced to move to the mainland. On the night of the earthquake, the unluckiest village of all, San Juan de la Costa, was wiped off its island by a tsunami, with a loss of two hundred souls.

On a local scale around the world, the sea level experiences changes for much less dramatic reasons as well. Tides cause the sea to rise and fall daily. Extraction of oil and water causes the land to sink on the Mississippi Delta, along the North Sea, on the Nile Delta in Egypt, and on the Niger Delta in Nigeria. During storms, winds blowing onshore combine with low atmospheric pressures to cause the sea to rise for a few hours or days. Seasonal or storm winds blowing offshore lower the sea, and during rainy seasons swollen rivers may rush to the sea and raise its local level. Major oceanic surface flows such as the equatorial currents pile up water on one side of continents and lower sea level on the other side. The level of the sea is about a foot higher on the Pacific side of the Panama Canal than on the Atlantic side because of a combination of lower water density and weather conditions on the Pacific side.

The geologic record shows that since the time that the glaciers reached their most recent maximum extent, 20,000 years ago, eustatic sea level rose 350 feet in mid-latitudes as the ice melted. Between 15,000

Figure 4.2 During the 1964 Good Friday earthquake, this shoreline segment along the Kenai Peninsula in Alaska dropped at least three feet. As a consequence of the resulting sea-level rise, the protective beach narrowed and the previously stable bluff began to be attacked by waves, a process that provides new sand to the beach. After a few more decades the new sand will have widened the beach to the point that the bluff erosion will halt (for a while) and trees will grow again on the bluff face.

and 6,000 years ago, the rise rate was 3 feet per century, although there may have been brief "blips on the curve" as high as 10 feet per century. Between 6,000 and 3,000 years ago, the overall rate of rise was around 1.5 feet per century. For the last 3,000 years the rate has been less than 4 inches per century. One hundred years ago the rate of sea-level rise "suddenly" accelerated up again, to between 1 and 1.5 feet per century. Adding the sinking of the Louisiana seafloor to the extant rising of the sea leads to a combined rise of 4 feet per century on the Mississippi Delta. Although it is widely expected that the rise rate will soon increase, there is as yet no indication that such is happening.

The devil is in the details. Coastal geologists are engaged in a raging controversy regarding the last 6,000 years of sea-level history. There is much geologic evidence of oscillations in the sea level, or "blips on the curve" during that time. The unresolved questions relate to how large the blips are and whether or not they are global in nature. The greatest blip may have been in the Southern Hemisphere along the shores of eastern South America, along all of Australia's coasts, and along both the east and west coasts of southern Africa. There, extensive evidence exists for a stranded shoreline, perhaps 3,000 or 4,000 years old, representing a sea level that was 3 to 6 feet higher than the current one. In northern Brazil, the sea level at that time may have been a full 15 feet higher than at present. Local variability in sea-level change is one reason why scientific information for coastal management purposes may best be viewed in a strictly local context rather than the global context that world governments support at present.

How can the sea-level history of the east coast of the United States be so different from Brazil's east coast during the last 6,000 years? After all, it's all the same ocean. One explanation for Northern and Southern Hemisphere sea-level differences may lie in the slightly distorted shape of the globe, caused by the wobbly rotation of a not quite perfectly spherical Earth.

Controversy over the degree and extent of it aside, sea level is rising along most tectonically stable coasts of the world, and along with that rise we can expect that storm waves and floods will reach higher levels and more inland locations than ever before. This is nothing new so far as Mother Earth is concerned. The sea has always been rising and falling, and global climates have always been changing. But it's new for us.

It is new because humans have lined the edge of the expanding oceans with cities, towns, and beach cottages. What would have been merely a predictable geologic event, easily absorbed by a flexible and malleable landscape, is today a potential catastrophe for an intractable and rigid modern humankind that crowds together at the shore. The great irony is that we humans are ourselves at least partly responsible for the sea-level rise that will eventually damage our own civilization.

How do we know the rate of sea-level rise has increased in the last century? For one thing, we have tide gauge records indicating that this change has occurred in many mid-latitude locations. In recent years it has become possible to measure sea-level changes by satellite. Nature also provides evidence in the form of barrier island thinning. A number of these Northern Hemisphere islands that are undeveloped, after having gone through a couple of thousand years of widening—or at least having undergone relatively small changes, are today rapidly thinning because of erosion on both sides. This is what's happening on some islands on the Outer Banks of North Carolina and on the German Frisian Islands. Shackleford Banks, North Carolina, is probably thinning by 10 to 20 feet per year, with the highest rates of erosion on the lagoon side, a process that most likely began in the twentieth century.

Some have cast sea-level rise as a civil rights issue rather than an environmental problem. In the December 2004 UN conference on global warming in Buenos Aires, representatives of the Inuit Eskimos from most countries bordering the Arctic Ocean announced plans to sue the U.S. government or U.S. industries or both for disruption of their lifestyle caused by the shrinking sea ice that allows generation of bigger summer waves during a lengthening summer. Joining the effort may be Pacific Island atoll nations such as the Marshall Islands and Micronesia, whose very existence is threatened by rising sea levels. In a May 2005 op-ed piece in the *New York Times*, Sujatha Bryavan and Sudhir Rajan predict that by the year 2080, up to 200 million people will be forced to flee rising sea levels. They propose that the refugees should be apportioned out to the top greenhouse gas emitters, with the largest number gaining entry to the United States.

For some atoll islands the future is already here. In November 2005 a milestone of sorts was achieved when the 980 inhabitants of the Carteret Atoll began moving to nearby Bouganville. Over the last twenty years the islands have been shrinking and saltwater intrusion is wiping out gardens and coconut palms. Local officials predict that the atolls will be submerged by 2015.

The 11,000 inhabitants of the nine Pacific atolls making up Tuvalu have arranged to immigrate to New Zealand over the coming years because of the threat of sea-level rise. In the Marshall Islands crops are being cultivated in old oil drums to avoid salt-soaked soils.

Modeling Global Sea-Level Rise

Climate variations will cause most of the changes in the level of the sea in the coming decades and centuries. According to Webster, *climate* is the average course or condition of the weather at a place over a period of years as exhibited by temperature, wind velocity, and precipitation. *Weather* is the state of the atmosphere at any given moment. Prediction of weather and climate are two entirely different things. Every day we are given weather forecasts, which are true quantitative predictions in every sense. They involve a complex natural system. Weather is predicted for the future, out there for all to see and to praise or damn. Even so, the best weather forecasts become uncertain in two to three days and are undependable six to seven days out. Weather has a short *persistence time*, which is the length of time for which weather can be expected to go along with change that is well known, understood, and predictable.

Prediction of global climate changes are important because of their possible impact on agriculture, fisheries, tourism, invasive plants and animals, natural hazards, quality of life, and, of course, the sea-level rise. The current most widely accepted prediction of sea-level rise is that its rate will be two to four times the present rate by the year 2100, and at that time the sea level will be two to three feet above its present state. At the same time, atmospheric temperatures will rise four to five degrees Fahrenheit. But these are numbers with a lot of leeway.

These are the predictions made by the Intergovernmental Panel on Climate Change (IPCC), a UN-related organization. The numbers are packed with vast uncertainties. They are vulnerable to criticism both by those who would maximize global climate change and its accompanying sea-level rise and by those who would deny it. Among the industries that are generally critical of global change science are the coal and petroleum industry, the auto industry, and the governments of beachfront communities. The resistance of these constituencies provided cover for the Bush administration in 2001 to refuse to honor the Kyoto Treaty, an international agreement intended to reduce the production of greenhouse gasses worldwide.

Ron Brunner, a University of Colorado policy scientist, characterizes the IPCC as the global change intelligence agency for world governments. Like any good intelligence agency, the IPCC is supposed to furnish the rationale for action or inaction, in this case on the problem of greenhouse gasses, and to suggest solutions to the problem of global warming. In the late 1980s, when the IPCC was first founded, its charge included the study of response strategies. As the global change issue became a political hot potato, world governments, fearing what "loose cannon" academics might propose to solve the problem, tucked the IPCC back into the fold of the pure science of global change and expunged its responsibility to study solutions. According to Brunner, world governments through the United Nations purposely designed the IPCC to minimize its usefulness and reduce its impact on society. IPCC studies of sea-level rise are global in nature and do not delve into local sea-level rise issues.

Understanding the future course of sea-level rise is arguably the most critical part of the intense international effort to study global climate change. Sea-level rise is already with us. At least a million people live within three or four feet of the level of the sea. Bangladesh, situated on the low-lying delta of the Ganges River, is the most striking example of a huge number of people immediately threatened by sea-level rise. The December 26, 2004, Indian Ocean tsunami, the greatest natural disaster in recorded human history, provided a disheartening measure of the concentration of human souls at low elevations next to Asian shorelines.

Sea-level rise is an issue of huge societal import. By some predictions, environmental refugees such as the people fleeing the lowlands of Bangladesh or the mid-Pacific atolls will far exceed the number of political refugees in the coming century. This assessment, which must be based on a conceptual model, perhaps should not be taken too absolutely, since it assumes knowledge of war magnitude in the next century!

The global change field of study employs a veritable army of scientists all over the world. The size of this effort is a credibility problem in itself. The United States alone is said to be spending in excess of $2 billion on climate research. Inevitably an "industry" of this size will seek to preserve itself and will not necessarily produce the simplest, quickest, and most powerful answers nor readily admit to fundamental problems blocking its way.

Numerous publications discussing both science projections and policy alternatives fill the shelves of the world's libraries. The *Climate Change 2001* report, 875 pages long and published by the IPCC, is the latest. A new volume is due in 2007. This report amply summarizes the scientific findings of climate change research, including the methodology of predicting climate change and sea-level rise.

Global change modelers usually employ a *bottom-up modeling* approach. Bottom-up modeling starts with the smallest elements of a system—a sand grain, a melting glacier, a reservoir, a cloud, a square mile of surface—and integrates all other processes and events over space and time to answer a question. This approach can include a long chain of events and a very complicated computer code, which is why global change modelers must use supercomputers. Assumptions are made at each stage along the way, and eventually, through a chain of calculations, the answer is reached. The problem in bottom-up modeling is that any errors in the assumptions and model weaknesses feed into the next layer or model, and the next, and the next, ever magnifying the initial error created by weak assumptions and uncertainties.

This approach assumes that if the physics of each process in the system is understood, the correct answer will be forthcoming. Bottom-up modeling approaches science with the same apparent degree of quantitative sophistication used in many branches of physics. The approach works well in physics and astronomy, so why not in earth science? The answer is that classical physicists study systems that are not complex and are far more predictable, such as radioactive decay, eclipse predictions, balls rolling down inclined planes, and so forth.

Top-down models start with the largest elements in the system, such as the Atlantic Ocean, the beach, the Greenland ice cap, or global atmospheric circulation, and hope to bypass the minutiae. The physics of the system are less important than its known behavior. In this approach, which relies more on field observations, the broad relationships between processes and events are sought. When painting with such broad brushes, however, numerous uncertainties are not addressed, leaving gaps that can be hazardous to the accuracy of model predictions.

Opponents of bottom-up models argue there are too many variables to handle, causing modelers to commit *the sin of commission*. Opponents of top-down models say too many important parameters are ignored by this approach, so modelers commit *the sin of omission*.

What a daunting task faces those who choose to predict the future of the sea-level rise! We have seen that the factors affecting the rate are numerous and not well understood. Even if our understanding improves, the global system simply defies accurate and quantitative prediction because of its complexity. Think back to the earlier explanation of the dynamics of sea-level rise. Some processes, such as the melting of glaciers, affect the sea level globally. Others, such as the depression and rebounding of the earth's crust, influence sea level only locally. As we build ever more dams on the rivers of the world, we use reservoirs to store huge volumes of water that would have gone to the sea and spread more water on millions of acres of irrigated fields. Some parameters, such as expansion of the sea caused by atmospheric warming, act slowly over long periods of time. Climate changes can cause sudden changes in ocean circulation, leading to sudden (decades) temperature changes that affect ice melting rates and cause sea-level change. At a very local level, pumping drinking water out of the sands that constitute a barrier island causes the grains to compact and hence a small local sea-level rise occurs. How do you account for all of these variables?

To digress, it is important to note that we are examining the quantitative models that will predict sea-level rise with sufficient accuracy to help society plan for the future. What to do about the Pacific Islands? Bangladesh? Manhattan? In order to be successful, the models must involve an accurate characterization of everything that plays a significant role in sea-level rise. We are not examining qualitative models that will tell us whether the sea level will likely rise or fall in the future. For these types of models we usually need only the most important parameters involved.

The big climate event that has become the standard fare of the cocktail circuit, the op-ed pages, and media talk shows is the *greenhouse effect*. Certain gasses, particularly carbon dioxide, produced in excess by the burning of coal and oil, accumulate in the atmosphere and prevent heat from escaping Earth's lower atmosphere; therefore, Earth retains that heat, much like a greenhouse. The carbon dioxide is said to be produced in excess by humans because it accumulates faster in the atmosphere than it can be absorbed by the oceans and by vegetation.

The most direct evidence of the increased CO_2 concentration in the atmosphere (and greenhouse warming) is provided by the work of Charles Keeling at the Mauna Loa Observatory in Hawaii. An atmospheric scientist from the Scripps Institution, he recorded CO_2 concentration for forty years at an elevation of two miles above sea level on the mountaintop. Compared to concentrations of 289 parts per million obtained from gas trapped in pre–Industrial Revolution glacial ice, Keeling's 1955 observations indicated CO_2 concentrations of 315 parts per million. Over the last forty years the concentration has steadily grown, to a maximum of 379 parts per million.

The evidence for the human connection to global warming is the correspondence of the massive production of excess carbon dioxide produced by burning fossil fuel in the last few decades and the simultaneous atmospheric increase in carbon dioxide. On a purely physical basis, the additional atmospheric CO_2 requires that some greenhouse warming must occur, but how much remains a question..

A number of important global climate change questions remain, including:

- How rapidly does the ocean absorb CO_2 from the air?
- How much excess CO_2 will humans produce in the future?
- How will global warming affect cloud cover?
- What will global warming do to ocean circulation?
- How will global climate change affect local climates?

Human behavior, the hardest variable of all to model, is central to the future of climate and sea-level prediction. Will we burn ever more coal and oil and continue to produce carbon dioxide, or will we come to rely on alternatives such as nuclear, wind power, or some other not even yet conceived source of power? How much carbon dioxide will China produce in the future? How much more water will we hold back from the sea through reservoir construction? How much more pavement will we construct, increasing runoff to the rivers and the sea? Global change models are looking 100 years into the future. How successful would such models have been if basic assumptions had been made in 1850 or 1900, before the automobile, the airplane, the power lawn mower?

University of Manitoba professor Vaclav Smil cites the unexpected change in CO_2 production in the 1990s as a good example of changes "whose timing and intensity and indeed the very occurrence" were not predicted by economic modelers. The combined fall of the Soviet Union and the rise of fuel efficiency in China simultaneously caused a reduction of CO_2 generation equivalent to its total global production in 1998 and 1999.

The uncertain future of Earth's ice cover provides another good lesson in feedbacks and complexity in mathematical models of sea-level rise. Ninety percent of the world's ice resides on the continent of Antarctica. The potential for sea-level rise tied up in Antarctic ice is huge. If all the ice on that continent melted, sea level would rise by 170 feet. The West Antarctic Ice Sheet, thought to be the most unstable part of the ice, would produce a 13-foot global sea-level rise upon melting. The Greenland ice cap has a 20-foot sea-level rise potential. One IPCC panel, however, predicted that during this century Antarctica could gain, not lose, ice because higher temperatures could lead to more snowfall. The prediction may be wrong. A 2006 report in *Science* magazine noted that during the previous three years the West Antarctic Ice Sheet has lost a surprisingly large 36 cubic miles of ice per year.

Melting of the Arctic Ocean ice cover (which has thinned by 40 percent in the last two decades) would not add to global sea-level change in any significant way, since it mostly floats on the surface.

The *albedo*, or reflectivity of solar radiation from ice and snow, is very high. Much of the energy of solar radiation is reflected from snow and ice, which leads to cooling of the atmosphere. On the other hand, the albedo of ocean water is very low. Thus, as warming creates open ocean water, especially during the brief Arctic summer, more and more heating of the water occurs, leading to more and more open water.

One of the most worrisome events that could occur as a result of the ongoing changes in the Arctic is the halting of the Gulf Stream. The Gulf Stream pushes warm surface waters north along the margin of the North American continent and eventually moves east across the North Atlantic, where it is responsible for keeping temperatures warm enough for

comfortable living over much of Northern Europe. There must be a counterflow of water to the south or else Gulf Stream waters would simply pile
up in the North Atlantic and the current would quickly come to a halt.
The counterflow requirement is satisfied by a large volume of water that
flows from the Arctic Ocean into the Atlantic. This water is cold and has
a higher-than-normal salt content, two attributes that make it heavy and
cause it to hug the seafloor on its journey to the south. It is cold because
it is formed under the ice cover of the Arctic Ocean, where the ice reflects
back a lot of the sun's energy. The high salt content is derived as salt is
released in the process of forming ice from seawater (sea ice is usually
fresh water). Today, with the warming of the Arctic, less salt-rich water is
released because less ice is forming, and the water is a bit warmer as the
solar energy is absorbed by the increasingly large areas of ice-free, open
water present in the summer.

It is likely that the counterflow of south-flowing water has already
been reduced in volume. This last happened during the so-called Younger
Dryas event 11,000 years ago. Because it is possible that the Gulf Stream
could be halted in a rapid time frame of just a few decades, Henry Pol-
lack, in his book *Uncertain Science . . . Uncertain World*, describes the
cessation of large-scale North Atlantic circulation as a hidden cliff edge.
We're probably close to the edge now, but no one knows where it is or
when we will step over it.

In one sense, the IPCC approach is a refreshing sort of modeling.
The publications of this diverse international group are filled with pain-
fully long discussions about errors, uncertainties, and missing data. The
objectivity of these global change modelers stands in stark contrast to
the arrogance of the coastal engineers or the overconfidence of groundwater modelers.

Objective or not, the problem of predicting sea-level rise is far from
solved, as is indicated in the final sentence in the 2001 IPCC report on
sea-level-rise prediction: "We recognize that it is important to assign
probabilities to [sea-level-rise] projections, but this requires a more critical and quantitative assessment of model uncertainties than is possible
at present."

But what is said in print in the middle of a dense 875-page document
and what is said out on the street are two different things. Somehow the
IPCC folds in the uncertainties (called "uncertainty absorption" by policy
scientist Ron Brunner), downplays the complexities, and comes out with
real predictions with error bars (pluses and minuses) of the future of
sea-level.

Table 4.1 Major Causes of Sea-Level Change

TECTONIC FACTORS—Unrelated to Human Activity
Sea-level rise or fall from tectonic movement—earthquakes and mantle convection
Sea-level rise from crustal depression (sinking of land) by the weight of ice or water
Sea-level fall by crustal rebound after the weight of a glacier is removed

GLOBAL WARMING FACTORS—At Least Partly Related to Human Activity
Sea-level rise from thermal expansion of ocean water caused by a warming atmosphere
Sea-level rise from melting of mountain glaciers caused by global warming
Sea-level rise from melting of Antarctic and Greenland ice caused by global warming
Sea-level rise from water released by permafrost melting in the high latitudes caused by global warming

TERRESTRIAL STORAGE ITEMS—Totally Human Related
Sea-level fall from water held from the sea by storage in reservoirs behind dams
Sea-level fall from reservoir water lost as it seeps into the groundwater
Sea-level fall from irrigation waters seeping into the groundwater
Sea-level rise from groundwater extraction
Sea-level rise from increased runoff from urbanization
Sea-level rise from deforestation

Source: IPCC 2001 Report.

Table 4.1 lists important causes of sea-level change from the IPCC 2001 report. They fall into three major categories: tectonic, global warming, and terrestrial water storage. Only the changes in the tectonic category are entirely unrelated to human activity.

The individual causes of sea-level change shown in table 4.1 are too numerous to discuss separately, but a cursory look at the so-called terrestrial storage terms provides an excellent example of model uncertainties. Although it is a minor player behind sea-level rise (relative to thermal expansion of seawater or the melting of Antarctic ice), it affords a good example of how the IPCC handles each of the parameters.

Terrestrial storage refers to the various means by which the normal paths for freshwater travel from land to the sea are altered. Water stored in the world's reservoirs and held back from the sea would seem to be a relatively straightforward parameter of sea-level change. But a number of questions remain about water impoundment, some of which are listed in table 4.2. Earth scientists Vivian Gornitz, of Columbia University, and Dork Sahagian, of the University of New Hampshire, seem to lead the way in evaluating these factors. That the two scientists often disagree is one measure of the tenuous nature of our understanding of the terrestrial storage terms.

As in most models of sea-level change, the value of uncertainty of each parameter is expressed as millimeters per year of potential sea-level rise. If a term in the model is expected to produce 1 millimeter per year sea-level rise, it will produce a 100-millimeter rise (approximately 4 inches) in a century.

Water is lost from many of the world's freshwater reservoirs through irrigation and other human uses. Sahagian believes that the extracted water eventually reaches the ocean by evaporation and plant transpiration, thus adding to the sea-level rise. Gornitz and her coworkers, however, believe that most irrigation water infiltrates back into the ground, becomes part of the groundwater, and thus is lost from the ocean. She does believe that some irrigation water will evaporate and increase the atmospheric water content, and thus is a factor that reduces sea-level rise by holding more water back from the sea.

Gornitz estimates that in 1990 the total contribution to sea level of all the terrestrial storage parameters was a negative (sea-level fall) -0.2 to -0.5 millimeters per year rate. Sahagian suggests that in 1990 terrestrial water storage resulted in a positive sea-level change of +0.06 millimeters per year. Here are two experts on terrestrial water storage whose predictions don't fall within one another's range of uncertainties. They don't even agree on the direction of sea-level change, whether it's falling or rising, resulting from water storage on land.

Table 4.2 Uncertainties About Uncertainties: Uncertainties in the First Two Items of the Terrestrial Storage List in Table 4.1

The volume of water stored in major reservoirs around the world is not well known.
The volume of the many tens of thousands of small farm ponds and rice paddies is not included, a possible 50 percent (?) underestimation. In one estimate the Brazos River drainage basin in Texas was said to contain more than 40,000 farm ponds.
Some of the water that fills a reservoir will be lost as it infiltrates into the local groundwater and builds up the water table.
The volume of water held as groundwater surrounding the reservoir is assumed to be 1.2 times the volume of water in the reservoir, but who knows? The number is based on "typical" rock porosity. But what's typical?
In some number of years, infiltration into the surrounding rocks will cease, once the rocks are saturated. When will that happen?
Volumes of many of the largest reservoirs have been reduced for irrigation and other water uses. How much reduced?

When two experts such as Gornitz and Sahagian disagree, the difference between their conclusions is sometimes assumed to represent the range of error in a particular parameter. For example, in the IPCC report, groundwater extraction during the twentieth century has a sea-level rise potential that ranges from 0.0 millimeter per year to 0.5 millimeter per year. This range comes directly from Gornitz's belief that groundwater, once used, returns back underground into the aquifer, hence the 0.0 millimeter per year minimum. Sahagian, on the other hand, states that most groundwater extracted from the earth eventually makes it to the ocean, hence the 0.5 millimeter per year maximum value.

The impact on sea-level change by terrestrial water storage that includes global groundwater extraction (table 4.1) is based on simplifications of physical processes, or *model simplifications*. In the process of arriving at a sea-level rise prediction, many such simplifications are made. But uncertainty determined by the distance between two scientific views is questionable, to say the least. This seems to be all the more reason why the final sea-level-rise prediction should be expressed in qualitative fashion.

The IPCC report concludes that the effect of the changing volume of water storage on the continents may be substantial and is increasing with time. The effect could be either a sea-level rise or a sea-level drop. We're not sure which!

Today there are probably fifteen major climate models in use around the world by various climate groups. In the IPCC volume alone, twelve models are used to determine future sea-level rise from thermal expansion of the ocean, and seven are used to project global average sea-level rise in the future. Effectively, tables 4.1 and 4.2 represent separate levels of models that are part of a pyramid of models something like that described in the Yucca Mountain discussion. Each model has its own set of unknowns and uncertainties, which feed into the next model and the next and the next. At the top of the pyramid are the models that put it all together.

Assumption upon assumption, uncertainty upon uncertainty, and simplification upon simplification are combined to give an ultimate and inevitably shaky answer, which is then scaled up beyond the persistence time to make long-term predictions of the future of sea-level rise. Aside from the frailty of assumptions, there remains ordering complexity: the lack of understanding of the timing and intensity of each variable.

One of the long-held criticisms of quantitative mathematical models in many fields is the use of corrections, referred to various names such as "coefficients," "adjustments," and "constants." These bring the results into line with perceived reality or into the "expected universe" conceived by Charles Perrow in his book *Normal Accidents*. Too often these corrections are simply fudge factors. Heat exchange between the ocean and the atmosphere is a very important model parameter, for both sea level and climate change studies. Modelers sometimes impose a "flux adjustment" to bring the heat exchange number into an expected range (the expected universe) of values. Steve Rayner, an Oxford University professor, describes flux adjustment as a "guesstimate" and "a clear example of how expert judgment (as opposed to the physics of the process) may play an important role in the modeling process" that is invisible to the policymakers and decision makers who use the results.

Raynor suggests that a second example of an institutional fudge factor is the anticipated change in global temperatures that results from a doubling of atmospheric carbon dioxide. Astoundingly, the range 1.5 to 4.5 degrees Centigrade is not empirically, experimentally, or model derived but is "the result of diffuse, expert judgment and negotiation among climate modelers."

When predicted numbers for sea-level rise or global temperature reach politicians, policymakers, and other users, they are usually unencumbered by any mention of the assumptions, uncertainties, and simplifications. Steve Raynor notes that "careful caveats about the scope and purpose of the models tend to melt into the background when both practitioners and users confront the apparent but misplaced concreteness of tables and graphs representing various model runs."

The objectivity of the IPCC documents is laudable. But the fact that the group recognizes its model weaknesses and is trying to improve them doesn't make its conclusions stronger or more believable. Besides, all is not peace within the IPCC family. In 2005, hurricane expert Chris Landsea resigned from the IPCC, saying that the panel writing up the section on the relationship between hurricane activity and global warming was contaminated by unsound science and a preconceived agenda.

Accurate prediction of future sea-level change is clearly impossible, but predicting the direction and the general magnitude of changes in the level of the sea is within the realm of our capabilities. General magnitude might mean prediction of a small rise rate increase defined as one foot per century or a large rate increase of five feet per century or perhaps a turnaround sea-level drop. Qualitative global change models can play a very important role here. The models have encapsulated all the major mechanisms, processes, and uncertainties that we know about, and have clearly indicated that the case for future sea-level rise is very plausible, however imprecise. But even such broad general predictions must be accompanied by the recognition that some of the unknowns are very large. For example, one scenario holds that as the ice on the surface of the Arctic Ocean melts, increased exposure will lead to increased precipitation of snow, which could lead to a sea-level drop.

We believe that global change modelers fall into two categories. There are the true believers who take no prisoners, believe every word, every model prediction, and feel that criticism is unwarranted or even un-American. A much larger group is uncomfortably aware of the insurmountable nature of the complexities in global change models. Pressed hard to the wall, this group admits the unlikelihood of providing the accurate prediction that society has grown to expect. However, the juggernaut, the large industry that has risen to answer the questions about global climate change, global warming, sea level rise, and all their ramifications, has unstoppable momentum.

The logical next step should be to turn toward more data-rich, qualitive modeling and to seek answers of a more general nature, to seek likely trends for the future, to example all the possible scenarios, the worst and best cases. It would make sense to spend a higher proportion of effort and money to gather field data to answer the many remaining basic questions about the future of the atmosphere and ocean. But the leaders in global change studies tend to view as a primary task the maintenance of funding for the modeling juggernaut. In this effort they are no different from most of the long line of supplicants who appear before Congress to ask for money to solve some social ill or technical or military problem.

Senator Inhofe's Fig Leaf

Never before in history has there been such an immense and concerted effort on the part of the scientific community to warn of an impending natural event. Besides the statements and studies of individual scientists, most major scientific associations have issued warnings about sea-level rise and climate change, the latest a 2004 statement from the 35,000-member American Geophysical Union. To the great dismay of most of the organization's more scientifically oriented members, the 31,000-member American Association of Petroleum Geologists gave its 2006 journalism award to Michael Crichton for his global warming–trashing novel, *State of Fear*. Of course, science isn't necessarily validated by a

Figure 4.3 Direct evidence of sea-level rise is furnished by the changed position of the barnacles between 1949 and 1981 on this bridge abutment near Miami, Florida. Photo furnished by Hal Wanless.

majority vote, and there is some variety of scientific opinion concerning the environmental impact of warming and the importance of humans as a cause of global warming.

This huge outpouring of scientific voices, however, did not deter Senator Inhofe. In fact, he could never have made the assertion that global warming may be "the greatest hoax ever played on the American people" without some cover, without some allies, without some comrades in arms. And there are plenty, some sincere and some motivated by powerful interests in this country (petroleum, coal, auto, power, and other industries) that are striving to maintain the status quo. The latter group will continue to downplay evidence of climate change, especially the role of excess carbon dioxide and the existence of global sea-level change.

It is easy to understand the basis of skepticism about the human role in global change because what's happening now has always happened. In the geologic past, sea level has always been changing, sometimes at much faster rates and sometimes at slower rates than at present. Global temperatures and rainfall rates have been higher and lower. Deserts have changed into tropical rain forests and glaciers have come and gone.

What's new is that the human race is here. If the scientific community is correct, we are causing global climate changes to occur rapidly and in a particular direction. If sea level rises (figure 4.3), it will affect millions who live near the shore. If the atmospheric CO_2 content rises and causes temperature and rainfall patterns to change, civilization will be profoundly affected. A debate about societal priorities in this rapidly evolving climate is needed.

We believe, however, that the senator's fig leaf would have been minuscule if the global change modeling community would firmly and publicly recognize that its efforts to truly quantify the future are an academic exercise and that existing field data on atmospheric temperatures, melting glaciers (figure 4.4), and other evidence should be relied on to a much greater degree to convince politicians that we have a problem. Let the models point to trends and answer the "what if" questions. A serious societal debate about "solutions" can never occur so long as modelers hold out the probability, just around the corner, of accurate projections of future climates and sea-level position.

In 2004 the National Science Foundation (NSF) and the National Center for Atmospheric Research in Boulder, Colorado, jointly announced the release of a new version of the premier global climate change model. Jay Fein, director of the NSF's climate program, described it as a significant milestone in climate modeling. William Collins, the chief scientist behind the model, is quoted as saying that it "has done remarkably well in reproducing the climate of the last century and we're now ready to . . . study . . . the climate of the next century." Tempting as it is to assume otherwise, the success of models in reproducing the past has little bearing on their success in predicting the future of a complex process on the earth's surface.

And once again a new model is announced, with implied assurances that future, usefully accurate predictions are just around the corner. "Give us another decade of funding and we'll tell you what to expect from global warming" is the unstated message of the bright, cheerful, and buoyantly optimistic press release. But it won't happen; it can't happen.

Often the language of the opposition is as virulent as Senator Inhofe's. In early 2004 an international group of nineteen scientists announced that their mathematical model studies indicated that more than a million land animal species would disappear from the earth by 2050. Perhaps a third of all life-forms are threatened, they believe. It's a complex thing. For example, some bird species are expected to disappear because their migration habits are timed to the seasonal appearance of insects that they prey on. But global warming is changing the timing of

Figure 4.4 A and B Comparison of the positions of the head of Muir Glacier, Alaska, showing the large retreat of the ice between 1941 (A) and 2004 (B). Widespread melting and retreat of temperate zone glaciers is a sure sign of global warming. The photos were taken from the same location. Photos furnished by Bruce Molnia of the U.S. Geological Survey.

insect life patterns and birds may not be able to adjust to the changed arrival time of their insect prey. Whether it's quantitatively modeled or not, and whether the impact would be nearly as massive as predicted, the study correctly identifies a major threat from global climate change.

Response from the opposing camp was rapid. Before the article actually appeared in the journal *Nature*, the Competitive Enterprise Institute, which promotes free enterprise and limited government, had its answer: "As in the past, when the agenda of the global activist elite suffers, improbable and speculative predictions of doom are trumpeted to terrify the world into compliance. With setbacks for boosters of global warming theory, like Russia's rejection of the Kyoto protocol, it's likely that they will begin clinging all the more tenaciously to ridiculous fantasies like this one."

A skeptical administration in Washington has also helped to provide context for Senator Inhofe's statement, starting with the rejection of the Kyoto protocol on global warming. The administration's opposition to the global accord was reinforced at the December 2004 United Nations conference on global climate change, held in Buenos Aires. The U.S. delegation rejected the European Union's suggestions of organizing a series of seminars to explore global warming issues and to begin looking at Kyoto's successor by vetoing any written or oral reports from such meetings. The U.S. delegation also stated that only one meeting should be allowed, the meeting must be for one day only, and issues concerning the future must be avoided. In December 2005 the American delegation walked out of a Kyoto Treaty meeting in a dispute over wording concerning the purpose of a future meeting. These were profoundly anti-intellectual, anti-science moves that received little attention from American media. It was also a hardening in U.S. policy toward the Kyoto protocol, moving from nonparticipation to active discouragement, even sabotage, of the international treaty effort.

In mid-June 2003 the *New York Times* reported its discovery that the Bush administration eliminated a long section of the Environmental Protection Agency's *Report on the State of the Environment* that described hazards from global warming and whittled it down to "a few non-committal paragraphs." They also replaced a reference to a study showing that temperatures had risen more rapidly in the last century than in the last 1,000 years with an American Petroleum Institute (API) statement that cast doubt on that conclusion.

The API is among the more powerful industry opponents of global change studies. The API, the chief lobbying and PR arm of the oil industries, says, "There is no credible evidence that the sea level is rising world wide as a result of human activities. However, changes do occur frequently from decade to decade or by region. For example, waters around the Mississippi Delta have been slowly rising but parts of Scandinavia have experienced a decline—not a rise—in sea level."

It is a clumsy and disingenuous statement. No objective scientist seriously questions the vast amount of data pointing to present-day sea-level rise on a global scale. Changes in sea level do occur at different rates and directions in local situations. For example, the Scandinavian sea-level drop mentioned in the API statement is well established as a local event, caused by rebounding of the land surface after the removal of the weight of the ice sheet.

Michael Crichton's negative view of quantitative mathematical modeling is very similar to ours, but we find ourselves in disagreement with his skepticism about global warming and sea-level rise. His 2004 novel, *State of Fear*, has generated a huge amount of controversy. TV advertisements for the book trumpet "They lied to you," an apparent reference to global change scientists and the establishment behind them.

In an appendix, Crichton compares the Nazi enthusiasm before World War II over eugenics and the power of Stalin's favorite genetic biologist, Trofim Lysenko (actually he was a non-scientist, peasant, plant breeder) with the current craze over global warming. Both eugenics (the basis of the Holocaust) and Lysenko's anti-Mendelian genetics (opponents of which were sent to the Gulag) became highly politicized, as has global warming. Apparently Crichton's point is that politicized science is bad science. The huge difference today, however, is that most in the geological, biological, oceanographic, and atmospheric science communities strongly support (in a nonpolitical sense) the reality of global change and a human role in it. The *New York Times*'s January 30, 2005, review of Crichton's book concludes that "this fellow has lost all perspective."

Politicization is inevitable for a theory that gores so many oxen of the high and mighty. Politicization, however, doesn't necessarily invalidate the science, as Michael Crichton seems to imply. But it does complicate things.

The opposition motivated entirely by economic considerations (coal, oil, and so on) is rarely seeking the truth; instead it is concerned with trashing the relationship between global climate change and excess CO_2 production. It is part of what some consider a new U.S. industry: manufactured doubt. The game is played at considerable monetary cost, and the goal is to cast doubt on any and all important scientific evidence favoring global warming. Favorable evidence is discarded, downplayed, or ignored; minor flaws become fatal flaws; and uncertainties become mortal weaknesses.

The scientific debate among those who are seeking only the truth is hampered in a number of ways when things become politicized. The lack of scientific integrity of the economically motivated opposition has a strong dampening effect on legitimate disbelievers or skeptics.

Scientists become defensive and may muffle or downplay criticism of certain evidence for fear that the opponents will blow the criticism out of proportion or that they will appear to be siding with the disingenuous opposition. The usual sharing of data between scientists dwindles for fear that others may distort the evidence in favor of one side or the other. The bandwagon mentality, always a problem in science, creeps in, making it difficult to stray from the beaten path.

Climatologist Patrick Michaels is a senior fellow of the Cato Institute, a think tank funded by oil and coal companies and tobacco, among others. Michaels, a University of Virginia faculty member, has a field day throwing out both the models and the concept of global change. How far the credibility of global climate change has fallen was amply illustrated on August 19, 2002, when Michaels was the "Newsmaker of the Week" on CNN's *Capital Gang* panel show. Panelist Katie O'Beirne asked if we were getting our money's worth for the $2 billion plus that we annually spend on climate change research, and Michaels responded, "Our computer models are really not much better than they were ten years ago before we started to spend all this money. . . . A new report that is being used to generate an awful lot of policy [in Congress] . . . was based upon two computer models that did worse [made worse predictions] than a table of random numbers when applied to U.S. temperatures [over the last hundred years]. . . . I think it is the biggest scandal in the history of environmental sciences." Panelist Robert Novak opined, "This is a plot . . . a conspiracy by people who don't like people who drive Corvettes . . . who drive SUVs." All in all, it was a bad day for the global change models.

Others ask why we should worry about the impact of sea-level rise. The models that predict future sea-level rise may well be wrong and sea level will soon be dropping, a frequent (and ridiculous) assertion of the American Shore and Beach Preservation Association, a group that promotes spending of taxpayers' money for beach nourishment to protect the huge summer rental industry of beachfront houses. The axe they grind on the sea-level issue is that if the public understands that the sea is expected to continue to rise, people may realize that it makes no sense to keep throwing dollars and sand into the ocean by nourishing beaches over and over.

Opposition to global warming has even become an issue with the Christian right. The Institute for Creation Research argues that the Bible shows there are global stabilizing factors that will minimize any global change that results from the greenhouse effect. Evangelist Jerry

Falwell intoned on CNN, "The whole [global warming issue] thing is created to destroy America's free enterprise system and our economic stability . . . but I can tell you that our grandchildren will laugh at those that predicted global warming. We'll be cooler by then, if the Lord hasn't returned."

Putting it all together, Senator Inhofe has a lot of friends and allies. His fig leaf is ample.

There is one great difficulty with a good hypothesis. When it is completed and rounded, the corners smoothed, the content cohesive and coherent; it is likely to become a thing in itself, a work of art. . . . One hates to disturb it. Even if subsequent information should shoot a hole in it, one hates to tear it down because it once was beautiful and whole.

—*John Steinbeck and Ed Ricketts*, The Sea of Cortez

chapter five

following a wayward rule

Shoreline Erosion on Sandy Coasts

The first high rise on the shoreline of the Outer Banks of North Carolina was built at Whalebone Junction in Nags Head, looming seven stories tall and situated well back from the beach. First a Ramada Inn, it later became the Armada Inn, accomplished by reversing the R and A in the original sign. Today it is a Comfort Inn, minus its restaurant, which was whisked away by storm waves. In a time frame of three decades, the shoreline marched up to the building until a corner of the hotel now sits on the beach extending to the mid-tide line.

So what do we do now? It is a question that is being asked along hundreds of miles of North American shoreline. The Whalebone Junction Comfort Inn is merely a microcosm of a global problem.

The steady landward retreat of the edges of continents, otherwise known as shoreline erosion, is perhaps the first significant negative global impact of the greenhouse effect. Perhaps the huge changes in the ice pack of the Arctic might also be considered a first impact. Although shoreline erosion is caused by many things, it is a certainty that sea-level rise is a major contributor. It is equally certain that with time, it will become an ever more important driving force of erosion.

The scientifically preferred term to describe a shoreline moving back in a landward direction is *shoreline retreat*, but the term *shoreline erosion* finds wide general usage, and we shall employ it here. In a strict sense, *erosion* is an inappropriate term because it implies loss or removal of material to some other location. A good part of the sand on a beach, rather than being lost, may move right along with the retreating shoreline. An eroding beach is not like an eroding slope of a mountain, where grains of sand, loosened by weathering, are removed by streams and eventually carried to the sea.

How we respond to this problem will likely foreshadow our response to other looming changes that will result from global warming. Will we take the long view and respond flexibly, or will we take the short view and defend the status quo? A flexible response means retreating from the shoreline—moving buildings back or letting them fall in. Flexibility in a democratic society also requires unusual courage and foresight on the part of politicians.

Defending the status quo next to the shoreline requires the citizenry to try and hold the margin of the sea in place with seawalls. This approach benefits the few who have built houses next to an eroding shoreline, but the long-term result (over several decades) is beach degradation for future generations and the loss of an entire ecosystem—birds, fish, crabs, and the critters they feed on. Alternatively, artificial beaches may be repeatedly pumped in at a cost of millions of dollars per shoreline mile.

It is important, right at the start of any discussion of shoreline erosion, to distinguish between *erosion* and an *erosion problem*. When a shoreline erodes and there are buildings or roads constructed next to it, there is clearly an erosion problem (figure 5.1). Absent these buildings, there may be erosion but clearly not an erosion problem. The beaches on Ossabaw, Wassaw, and St. Catherines islands in Georgia and Hunting Island in South Carolina are littered with fallen trees, a sure sign of erosion. But with no buildings near these shorelines, there is no erosion problem, and in fact the log-covered beaches are hauntingly beautiful. Shoreline erosion is a natural response to rising sea level and a host of other events, and thus it is a problem only when causing an inconvenience to humans by threatening their structures.

Eighty percent to 90 percent of North American ocean shorelines are eroding, and probably the same is true for those in the rest of the world. Rates of erosion range from tens of feet per year, as on some

Figure 5.1 A 2005 erosion problem at the Wild Dunes Development on Isle of Palms, north of Charleston, South Carolina. The meager dunes shown here are hardly wild. They are the remnants of a bulldozed pile of sand taken from the beach, which in itself is another form of shoreline erosion similar to beach mining. Photo furnished by Nancy Vinson.

islands along the sinking Mississippi Delta, to less than a foot per year along some parts of the west coast of Florida. Extreme rates of erosion of feet per day may occur during big storms that strike sandy coasts, but much of this land loss may be recovered in the days, weeks, and months following the storms as sand is pushed ashore by fair-weather waves. Irreversible loss rates of feet per day happen when beaches subside during earthquakes, as in 1964 along Alaska's Kenai Peninsula and in 1972 along the tropical Pacific coast of Colombia.

Present-day shoreline erosion rates around the United States coastal plains on the East and Gulf coasts usually fall between two and four feet per year. Certainly a part of this erosion is related to sea-level rise, but at present we do not know how to sort out the relative importance of all the various natural causes of shoreline retreat. Adding to the complexity of the problem of erosion rate prediction is the role of humans, who build dams on rivers, dredge navigation channels, construct jetties and seawalls, and mine sand on beaches. All these actions cause loss of sand to the beach, and a diminished supply of sand leads to increased rates of erosion.

Right now in the United States, sea-level rise is mainly a threat to beachfront development in the hundreds of lower coastal plain and barrier island recreational communities. From the south shore of Long Island, New York, to Padre Island, Texas, there extends a seldom-broken chain of eroding beaches that threaten seashore communities. On a longer time frame of three to five generations, the retreating shoreline and rising seas will be a problem for lower coastal plain cities all over the world. In the United States, this includes cities such as Galveston, New Orleans, Tampa, Miami, Charleston, Philadelphia, and Atlantic City. Even New York City and Boston, while, strictly speaking, are not located on coastal plains, are nevertheless at low enough elevations to be threatened by rising seas.

It is here, along the world's low-lying, flat coastal plains, that sea-level rise is most important in the immediate future. Coastal plains are low, broad plains with little relief that extend landward from an ocean shore to the nearest elevated land. In Georgia the coastal plain extends from the shoreline to Atlanta. In North Carolina it extends landward to Raleigh. The Gulf of Mexico and Atlantic U.S. coastal plains are aprons of gently seaward-dipping sediment layers derived from the erosion of ancient mountain chains, such as the Appalachians.

The slope of coastal plains is so slight that a very small sea-level rise has the potential to cause a massive landward retreat of the shoreline. For example, the average slope of the lowermost coastal plain in North Carolina is about 1 to 2,000. This means that for every 2,000 feet of distance landward from the beach, there is a one-foot rise in land elevation. It also means that the theoretical shoreline retreat, assuming that only the slope of the land controls a response to sea-level rise, should be on the order of 2,000 feet per foot of sea-level rise.

By contrast, the land is so steep along our mountainous Pacific shores that a small sea-level rise won't make much difference in shoreline erosion rates. Of more importance in causing shoreline erosion along these steep coasts is the damming of rivers, which shuts down the annual contribution of sand from the spring floods.

If sea level were to suddenly jump up twenty feet tomorrow, there is no question where the new shoreline would be. It would be at what is now the twenty-foot contour line on maps. That's obvious. But that's not the question facing coastal communities today. The question is, What

Figure 5.2 Construction workers celebrate the completion of a condominium by fishing in the surf zone of Garden City, South Carolina (circa 1986). There is a distinction between erosion and an erosion problem. Here the building jammed up against a shoreline has created an erosion problem. A good rule of thumb is you should not be able to fish from your condo windows. Photo furnished by the *Coastal Observer*.

will be the result of a minuscule rise of less than one-eighth of an inch per year?

The battle is almost lost in low-lying Bangladesh, on the Ganges-Brahmaputra River Delta, where there is nowhere to go, no higher land for escape. A flood of "environmental" refugees is a certainty in this century as millions who now live within three or four feet of sea level will be forced to flee to adjacent countries. Although there is a big problem in Bangladesh, in much of the rest of the developing world the shoreline erosion problem is much less worrisome. Money is rarely available to try and halt the inexorable landward march of shorelines on the barrier islands of Colombia, Northern Brazil, and Nigeria. So buildings are routinely moved back and villages are sometimes altogether abandoned to avoid the retreating shoreline. Locals live very flexibly with the situation.

Meanwhile, for the Western world the erosion problem gets visibly worse on an annual basis (figure 5.2). Bigger buildings that overshadow the Whalebone Junction Comfort Inn are built next to a shoreline that is steadily retreating toward the structures. In Gold Coast, Australia, an eighty-story condominium is found at low elevation next to an eroding shoreline. Development rushes to the shore just as surely as the shore rushes to the development.

But fear not. To the rescue have come the coastal engineers, dedicated to defending the status quo, to holding the shoreline in place indefinitely. But drawing a line in the sand hundreds of miles long and steadfastly defending that line is an economic impossibility in a time of rising sea level unless a nation is willing to dedicate resources to the task at the same level that the Dutch do.

Why Do Beaches Erode?

A shoreline exists in a state of equilibrium between the natural forces that move sand away from a beach and the forces that supply sand to the beach. Every beach has a different combination of forces involved, but waves do most of the work.

Sand is gained on a beach from the following sources:

- the continental shelf
- adjacent beaches
- rivers
- bluff and dune erosion

Sand is lost from the beach to the following "sinks":

- the continental shelf
- adjacent beaches
- sand pushed onto land by storm wave overwash
- sand blown up on land by wind

Putting it all together, the sand volume gain minus the sand volume loss equals the sand supply to the beach, a number that may vary considerably from year to year. If the beach gains sand, it will accrete. If the beach loses sand, it will erode.

Comparison of a beach in South Carolina with a beach in Iceland illustrates how this relationship can work. Hunting Island Sate Park, South Carolina, is retreating at the very rapid rate of tens of feet per year. But the waves that provide the energy to move sand and erode the beach are small. A large natural offshore sand shoal has trapped Hunting Island's sand supply, which comes from barrier island beaches to the north. Rivers are not contributing any new sand here, so the temporary blockage of sand has led to accelerated erosion and a beach covered with fallen trees.

By contrast, the beaches on the barrier islands of southeast Iceland are accreting, or advancing seaward, at about the same pace that the Hunting Island beach is retreating. Some of the highest waves in the world strike this portion of Iceland's coast, where forty-foot waves can be seen just offshore every winter. The Icelandic shoreline is accreting because a huge volume of sand is supplied by glacial meltwaters as nearby glaciers retreat (probably because of global warming). Particularly large surges of new sand arrive at the beach each time a volcano erupts under the ice.

Another very serious erosion problem threatens the equilibrium of the Arctic Inupiat Eskimo villages along the North Slope of Alaska and probably along parts of the Siberian and Canadian Arctic coasts as well. Normally the Arctic summer, featuring open, ice-free water between the permanent ice pack and the mainland shoreline, lasts two months, July and August. Starting in September, ice begins to re-form and the earliest small slabs of ice are broken up by the waves and pushed ashore. The ice, with incorporated beach sand, forms a frozen natural seawall that absorbs the waves of the storms that arrive in October. By November the sea is solidly frozen over and storms are no longer a threat to the shoreline. But the Arctic is warming, summer is lengthening, and the sea remains free of ice into October and sometimes into November. Now the October storms strike an unprotected shoreline. In addition, the longer, warmer summer increases the aerial extent of ice-free water off Arctic and Chukchi Sea villages. More open water leads to increased distance over which winds can generate waves, which in turn leads to larger waves.

Along some coasts, for example, those of Texas and California, rivers furnish a fresh supply of sand to the continental shelf after every flood. Once the sand is deposited on the continental shelf in front of the river mouth, some of the former river sand is then pushed up to the beach by fair-weather waves, a process often observed but poorly understood. In this fashion, sand may work its way back to the beach from water depths as great as forty or fifty feet. Sand is lost in the opposite direction as well, particularly during storms. Strong seaward-directed currents along the sloping seafloor transport sand from the surf zone well down the shoreface. In truly major events such as hurricanes and big, slow-moving winter storms, nearshore sand may be transported miles out to sea and may even cross the entire continental shelf to end up in deep water on the continental slope. Along the California coast, sand moved offshore often flows directly from the beaches into the heads of submarine canyons. From there it flows down the canyon axis to reside forever in the deep ocean waters.

Some streams along the mountainous Pacific Coast dump their sand directly on beaches. Along much of the East Coast, however, rivers dump their sand load tens of miles up the relatively still waters of estuaries at the river mouths, and little if any of the river sand makes it to the coast. In other words, along mountainous coasts, new sand arrives to the beaches every year (at least that was the case before the sand-trapping dams were built), but on coastal plains coasts there is no "new" sand.

Sea-level rise is a component of this dynamic equilibrium between sand supply and wave energy. As sea level rises, the storms and tides push farther inland, bit by bit, adding to the rate of shoreline retreat and to the frequency of overwash contribution of sand to the upland. As sea level rises, the biggest storm waves can be expected to hit California rock cliffs a few more minutes each year and eventually a few more days each year. Since sea level is expected to continue to rise and the rate of rise will probably accelerate in the foreseeable future, erosion rates should accelerate as well, a trend that will be helped along by expected increased storminess. These conclusions are based on qualitative modeling.

Some of the future erosion of shorelines caused by the sea-level rise will operate through convoluted, even tortuous paths. Research by Duncan Fitzgerald, a Boston University coastal geologist, indicates that changes in the hydrodynamics of lagoons behind barrier islands lead to erosion of beaches on the ocean sides of the islands. It works this way: The water level of the lagoons behind barrier islands will slowly rise apace with the sea-level rise. More water in the lagoons will lead to a larger water volume and stronger currents for the twice-daily tidal exchange with the sea through the inlets. Stronger tidal currents mean that the ebb tidal delta will enlarge. The ebb tidal delta is the body of sand that protrudes out to sea at every barrier island inlet. Such deltas

Figure 5.3 Mining a beach in Portugal creates an erosion problem, or at least intensifies it. All over the world, mining of beach sand is a major cause of shoreline erosion despite the fact that it is usually illegal. Mining, however, is one of many factors, besides sea-level rise, that affects the rate of shoreline retreat.

are responsible for uncounted shipwrecks and also provide the surface across which sand is exchanged between adjacent islands. The sand that is used to enlarge the tidal delta is the sand that normally is transferred from one island to the next island in a downdrift direction. Because sand is "stolen" to enlarge the tidal delta, the sand supply is reduced and the erosion rate on the adjacent island will increase.

Now add the human impact. We build dams across rivers and trap sand that belongs to the beaches. This is a big problem on the West Coast of the United States and a much smaller problem along the eastern U.S. barrier island coast. Humans also dredge navigation channels through inlets, thus keeping sand from crossing from island to island across the ebb tidal delta. And humans who are imprudent enough to own beachfront buildings build seawalls, along with groins and offshore breakwaters, all of which eventually reduce and even halt beach sand transport. Sand is bulldozed from the beach and piled up against buildings to protect them. Sometimes sand is outright mined from the beach to furnish sand for concrete buildings to be built next to the same beach (figure 5.3). In some developing countries, beach sand is the major source of construction sand. We once saw a front-end loader filling a dump truck with sand from the low-tide beach in front of the premier coastal resort of Ecuador. The beaches on many Caribbean islands, such as the Virgin Islands and Puerto Rico, are a mere shadow of their bulk of two hundred years ago because of a long history of humans' mining of the beaches for concrete.

Humans also bring politics and avarice into the shoreline erosion picture. Big money is involved, always a fatal blow to good science. Erosion rates form the basis of zoning and setback regulations, and they are part of the justification for federal support of beach replenishment projects. Erosion rates, or even the mere recognition of an erosion rate at all, are construed to be bad for the real estate business. The politicians of Dare County, North Carolina, opposed moving the Cape Hatteras Lighthouse even though it was sure to be lost in some future storm if it was not moved. The reason: the move would attract worldwide attention to the erosion problem of North Carolina's Outer Banks (which proved to be true).

Figuring out just how sea-level rise will affect this complex shoreline equilibrium of many parameters and how it will affect shoreline retreat on a particular beach is the most daunting and perhaps the most important task facing coastal scientists.

To date, a single mathematical model purports to predict shoreline retreat caused by the rising sea. Called the Bruun Rule, it is a very simple model that is used to describe a very complex process on sandy coasts. The use of the model is as global as the shoreline erosion problem itself.

The Bruun Rule and a New World

The Bruun Rule resides in a world dominated by engineers rather than scientists. It is a world where it is not possible to admit defeat and walk away or to respond flexibly, one where an answer must be found and where the answer, to be credible, is best found by the most sophisticated means possible. This is the world of coastal engineering mathematical modeling.

Per Bruun, a Danish American engineer whom many consider to be the father of modern coastal engineering, devised the concept of a shoreface profile of equilibrium in the mid-1950s. Within a decade, the rule evolved into a theory on how shorelines retreat in response to sea-level rise. This theory in turn evolved into a means of predicting the amount of horizontal retreat of the shoreline caused by sea-level rise. Maury

Schwartz, a coastal geologist from Western Washington University, gave it the name Bruun Rule to honor the inventor.

Bruun's view of the beach correctly included the entire *shoreface*, the narrow, concave, and relatively steep zone extending seaward from sandy beaches. His major contribution was to recognize that erosion was not merely the retreat of the wet-dry line on the beach but involved the landward shifting of the entire shoreface. In reality, the shoreface is the lower beach. Along most of the United States' East Coast, the shoreface is less than a mile wide and extends to a depth of thirty to sixty feet. Normally the shoreface is much steeper than both the mainland coastal plain and the seaward continental shelf. It is a surface that is easy to spot on navigation charts because there is an abrupt flattening of the seafloor where the true continental shelf begins. The shoreface is the zone within which beach sand oscillates back and forth according to wave conditions. The visible upper beach and the shoreface are constantly exchanging sand. During times of fair weather, sand on the shoreface tends to move toward shore. When storms strike, the direction of sand movement usually reverses.

The critical explicit assumption behind the Bruun Rule is that as the shoreline erodes and the sea level rises, the whole shoreface is displaced landward and upward in proportion to the sea-level rise (figure 5.4). It is assumed that the shape of the shoreface always remains the same, in a so-called profile of equilibrium. In order to maintain the equilibrium profile as the shoreface moves up and back, sediment is eroded from

The Bruun Rule of Shoreline Retreat

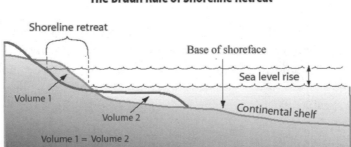

Figure 5.4 Diagram showing how the Bruun Rule works. As sea level rises, the shoreline moves back and up, while the profile of the shoreface remains the same. The volume of sand removed from the upper part of the profile is the same as the volume of sand deposited on the lower profile. In reality, the mechanics of shoreline erosion are very complex and the Bruun Rule is at best a special case. Drawing by Charles Pilkey.

the upper shoreface (and beach) and redeposited on the lower shoreface. The sediment eroded is roughly the same volume as the sediment deposited. Waves and waves alone move the sand back and forth.

The Bruun Rule assumes that the shoreface is made up entirely of sand that is of uniform grain size and that there is no net gain or loss of sand by longshore currents, wind action, or storm overwash. The assumption is that the slope of the shoreface is controlled entirely by the grain size of the sediment covering its surface—the coarser the sand, the steeper the shoreface. But in fact, the slope of the shoreface is determined by many factors, not just grain size. And furthermore, although a shoreface profile of equilibrium may exist, much more is involved in constructing it than just waves. It is clear from the last two decades of shoreface studies that the energy of the waves, the frequency of storms, the direction from which storms arrive, the size of the supply of sand to the beach, the underlying geology of shorefaces (rock or mud layers), and grain size all play a role in determining shoreface shape. There is no reason to expect, and no field evidence to support, the notion that the shoreface shape remains the same as the sea level rises.

Per Bruun's deceptively simple notion from the 1950s about the shoreface response to sea-level rise was an important advance in its time. It was a concept that held sway for a whole generation of coastal scientists. In a classic paper published in 1975, Don Swift, of Old Dominion University, agreed that the entire shoreface moved with sea-level rise. But he recognized that the shoreface would not retain a constant shape but would change according to storm magnitude, sand supply, and other factors. The Bruun Rule represented a single case in a broad spectrum of shoreface responses to sea-level rise. Next came Donn Wright, of the Virginia Institute of Marine Sciences, who put instruments on the shoreface that measured waves, currents, and sand transport. The net result was that the shoreface processes proved to be far more complex than either Swift or Bruun had envisioned. For example, bottom currents were much more important that once believed.

The next major bump has been the work of Rob Thieler and William Schwab, of the U.S. Geological Survey. Taking advantage of new technology in precise satellite navigation and sonar (sound) mapping of the seafloor, they produced startlingly accurate side-scan sonar "photography" of the seafloor using sound instead of light, maps showing frequent rock outcrops on the shoreface, and a number of surface features indicating heretofore unsuspected currents. Rock outcrops and currents aren't accounted for in the Bruun Rule.

Evidence continues to accumulate from all over the world that the basic assumptions behind the Bruun model are very wrong. Yet it continues to be widely applied by coastal scientists, who should know better, and blindly applied by social scientists, planners, and international agencies concerned with how future global trends will affect coastal cities. The problem is that there is no alternative to the model. The Bruun Rule is the only model that predicts shoreline erosion resulting from rising sea level. It is the only show in town.

In actual application of the Bruun Rule, all one has to know in order to predict the future rate of sea-level-caused erosion is the rate of sea-level rise and the slope of the shoreface. In practice, slope is determined using a navigation chart. When all is said and done, and the original equation with four or five variables is simplified, it boils down to the fact that the predicted rate of shoreline retreat is proportional only to sea-level rise and the slope of the shoreface (see appendix). The rule predicts that the steeper the shoreface, the slower the shoreline retreat. There is no basis to believe that slope controls erosion rate. In fact, it should be the reverse. Other things being equal, the steeper a purely sandy shoreface, the more rapid the shoreline retreat.

Hoover Mackin, a pioneering geologist who studied landforms, observed long ago that it is quite possible to say absurdly stupid things in the language of mathematics that would produce intense embarrassment to the speaker if communicated in plain words. The Bruun Rule certainly qualifies as an example of Mackin's observation.

Like all models and equations that predict a process on the surface of the earth, a number of implicit and explicit underlying assumptions are made in the Bruun Rule. Some of these assumptions are not at all apparent to the non-specialist who is relying on the Bruun Rule equation. Virtually none of these assumptions work. This may not have been apparent at the time Bruun came up with the idea, but today we know better. The following examples illustrate this point.

There is an offshore depth (called the closure depth) beyond which it is assumed that significant amounts of sediment from shallower water will not pass. Effectively, closure depth is a sand fence that blocks the seaward flow of beach sand in a storm. It has now been well established that during truly large storms, sand can be moved many miles off shore—ten miles, even fifty miles. The sand fence just isn't there, not even when the waves are small.

Sand movement on the shoreface is assumed to be caused only by waves. Bottom currents are unimportant. The closure depth concept was based on the assumption that only waves can transport sand and that as the water gets deeper, the bottom stirring action of waves eventually is reduced to nothing (at the closure depth). But today bottom currents of many origins, moving in virtually any direction, are very well documented. When combined with the stirring-up action of waves, they can carry large amounts of sand right through any sediment fence.

A thick sand layer without rocks, mud, or other complications is assumed to underlie shorefaces. Recent studies, mostly carried out in the search for sand to be mined to pump up on nearby beaches, have indicated that shorefaces are often, if not usually, underlain by some combination of hard rock and mud layers. Hard rock layers may be ancient (millions of years old) limestones and sandstones or they may be sand and shells cemented just a few thousand years ago. Naturally, these strongly influence the shape of the shoreface and also the rates of shoreline retreat. The beach atop rocky shorefaces such as the one off Myrtle Beach, South Carolina, can be expected to erode more slowly than one with former marsh mud outcroppings on it, such as on Cape Henlopen in Delaware, Parramore Island in Virginia, or the Brazos Peninsula in Texas. Erosion rates affected by the underlying geology of the shoreface may be entirely unrelated to rates of sea-level rise.

Clearly, retreat of the shoreline caused by sea-level rise is a complex process. Two different shoreline reaches (such as South Carolina and Iceland) undergoing identical gradual sea-level rises may erode at very different rates. This is best illustrated by a look at the parameters that possibly could be involved. Listed below are a number of processes or conditions that will accelerate or hold back shoreline retreat over a societally useful predictive time frame of 25 to perhaps 200 years. The list is applicable to sea-level-rise rates of one to three feet per century. With higher rates of sea-level rise and/or longer time frames of 1,000 years or more, slope of the land surface will be the principal determinant of shoreline position related to changes in the level of the sea. The rate of sea-level rise, which is determined by a combination of eustatic and isostatic processes, is the underlying cause of erosion. The actual movement of the shoreline, however, is accomplished by a variety of other processes. This is why different shorelines subjected to identical sea-level rise may erode at significantly different rates.

Primary Factors
Sea-level rise
Regional tectonic uplift or downwarping (e.g., in earthquakes)

Subsidence due to sediment compaction (e.g., from oil and water extraction)

Secondary Factors
Present-day erosion rate
Slope of the upland (the mainland)
Relative importance of inundation versus erosion
Coastal type (rocky, marshy, mountainous, barrier island, bay, open ocean)
Shoreface slope
Previous sea-level-rise history (is the shoreface slope adjusted to present conditions?)
Upland geology (unconsolidated sand, rock?)
Upland morphology (bluffs, dunes?)
Volume of sediment in the way of shoreline retreat (e.g., size of a barrier island)
Lagoon size
Geology underlying beach (rock type, mud, sand)
Beach and shoreface grain size
Wave climate
Fetch
Tide character
Width of continental shelf
Storm types and frequency
Storm response
Storm surge potential
Storm surge ebb potential
Sediment supply
Beach rock or other lithification processes
Shoreline vegetation

Impact of Humans on Sediment Supply
Rivers (dam construction or dam removal)
Dredging
Nourishment
Hard stabilization (jetties, groins, seawalls)
Climate change impact on sediment supply, storms, vegetation
Arctic Shorelines
Permafrost conditions
Time span of open water
Summer fetch

The list illustrates *ordering complexity*. This critical concept means that even if you know how each of the factors works and interacts with other factors, including sea-level rise, in causing shorelines to retreat, you still can't predict the future because you don't know the order in which the factors will occur. Each shoreline segment will have a unique combination of these factors responsible for its unique rate of shoreline retreat in response to sea-level rise. On different shorelines the various parameters will be of varying importance, over varying time frames.

This is ordering complexity. This is why shoreline retreat related to sea-level rise cannot ever be accurately predicted.

Who's Propping up the Rule?

The uncritical use of an outdated model such as the Bruun Rule, and the failure of the engineering and science communities to incorporate a world of new observations concerning shoreface evolution, border on the scandalous. Judging from the scientific literature, especially the coastal zone management literature, two principal proponents of the use of the Bruun Rule seem to be Stephen J. Leatherman and R. J. Nicholls. Leatherman, best known as "Dr. Beach," annually produces the list of the nation's top ten beaches. He is a professor at Florida International University in Miami and director of the International Hurricane Center. Nicholls is at the Flood Hazard Research Centre at Middlesex University in Enfield, UK.

These scientists have repeatedly recommended, defended, and promoted the Bruun Rule as a coastal management tool to predict the impact of future sea-level rise on shoreline erosion for many communities and countries. Along the way, they have made some strange claims. Leatherman and some coworkers suggested that sometimes tens or hundreds of kilometers of shoreline could be represented by a single cross-shore profile and therefore a single erosion rate, although no evidence to support this doubtful claim was presented. Nicholls and coworkers advocated the use of the Bruun formulation for sand and gravel coasts, cliffed coasts, and, with some reservations, muddy coasts, all highly doubtful possibilities. To their credit, Leatherman and his associates frequently recommend extrapolating present-day shoreline retreat rates into the future as a check on Bruun Rule results (model verification).

Leatherman and his coworkers claim to have demonstrated that the Bruun Rule does predict shoreline retreat caused by sea-level rise. As with any mathematical model, however, one needs to lift the flap of the tent and peek underneath before accepting any proof that the rule works. To prove the validity of the rule, they chose sites along portions of the New Jersey, Delaware, and Maryland coast where, supposedly, sea-level rise and nothing else was responsible for shoreline erosion and success-fully applied the rule. Their validation of the Bruun Rule was immediate-ly challenged by a number of scientists, the fundamental criticism being that there is no open ocean shoreline where only sea-level rise causes erosion. Applying the rule to a shoreline that is eroding from several causes provides no basis for testing the rule. How can you isolate the role of sea level? The critics also noted that the rule is an ill-founded concep-tual model with no observational basis.

In 2005 Nicholls, along with Dutch modeler M. J. Stive, declared that the Bruun model isn't wrong; it is just incomplete and needs to be "more comprehensive." That is, the inclusion of more variables will provide more accurate predictions of sea-level-caused shoreline retreat. But what variables from the long list above would you choose? And why didn't Nicholls note the fact that the rule doesn't quite work earlier—before it became globally applied for coastal management purposes?

The Politicization of the Bruun Rule

When Per Bruun devised his rule, he assumed that the closure depth or sand fence was well offshore, at a water depth of sixty to seventy feet off the east coast of Florida, for example. Probably sand escapes past a "fence" at that depth off Florida only during the biggest storms. With time, however, beach nourishment politics reared its ugly head, and the Florida closure depth became subject to change, moving to ever more shallow water, eventually to a depth of seventeen feet. The reason for the shoaling (on paper) of the sediment fence was that it made beach nourishment projects cheaper. According to standard beach design prin-ciples, the shallower the sediment fence, the smaller the amount of sand and the lower the cost of building an artificial beach.

There is virtually no field evidence to indicate that the closure depth off east Florida is at seventeen feet—or at any other depth, for that mat-ter. But there it is, ensconced firmly in the engineering literature and in U.S. Army Corps of Engineers documents on replenished beach design.

A second distortion of Per Bruun's original thinking concerns the control of the profile of equilibrium. We think Bruun realized that the shape of the offshore profile of the shoreface was determined by a lot of things in addition to grain size of the sand on the seafloor. As mentioned earlier, it is clear from the last two decades of shoreface studies that waves, storms, bottom currents, the supply of sand, the underlying geology, and grain size all play roles in determining shoreface shape. Today it is a rock-hard engineering principle for the purpose of mathematical modeling of beach behavior that grain size and grain size alone determines the profile shape. In other words, if one knows the grain size of the sand covering a shoreface anywhere in the world, the profile shape is known as well. All shorefaces with the same grain size have the same shape. It is such nonsense.

It also turns out that the choice of grain size is itself a subjective process. The typical shoreface along the East Coast of the United States may have a range of sand sizes, including shell gravel or mud outcropping on it. This situation leaves the coastal engineer basically free to choose whatever grain size he or she wants. Different grain sizes will give different shoreface slopes, and thus the Bruun Rule will provide different predicted erosion rates. Take your choice.

Reliance on a theoretical shoreface shape determined on the basis of grain size is a lot easier than getting out and measuring the real thing in a crashing surf zone. In the 2001 design of jetties at Oregon Inlet in North Carolina, the shape of the shoreface profile was needed as part of the quantitative mathematical modeling process of sand transport along the beach by waves. Numerous actually measured, real-world profiles were available from field studies near the inlet. But instead of using the real thing, the engineers determined the profile by the equation based solely on the size of the sand on the beach. The design engineer wrote that the real profiles were quite variable over time and place and this "complicated" the design process. This engineer found it much easier to have a single simple profile determined mathematically from a sand sample analysis than to have to account for all the muddled, tangled confusion and beauty of nature.

Who Uses the Rule?

All the world seems to embrace the Bruun Rule. We suspect that it may be the single most widely used mathematical model to predict the outcome of a natural process on the surface of the earth. Its heuristic appeal and simplicity are apparently irresistible. Buttressed by the name and reputation of one of the giants of coastal engineering, the rule has achieved international status. Recommended by government agencies in most of the world's coastal countries on six continents, as well as world bodies like the UN's Intergovernmental Panel on Climate Change (IPCC), it seems to have a rosy future, whether or not a solid scientific basis exists for the model.

Below is a partial list of countries known to have used the Bruun Rule in some fashion since 1995, in some aspect of coastal management. Interestingly, a number of the published coastal management documents from which the list was compiled recognize some of the weaknesses and criticisms of the rule but nonetheless recommend using it. The attitude seems to be, "It's the best model we have, it's the only model we have, so let's use it until something better comes along."

Argentina
Australia
Barbados
Brazil
Canada
East Asia
Eastern Caribbean
Egypt
Estonia
Granada
Holland
Indonesia
Japan
Lebanon
Malaysia
New Zealand
Nigeria
Pacific Islands
Senegal
South Africa
The Gambia
United Kingdom
United States
Uruguay
Venezuela

One result of the global use of the Bruun Rule is a frequent mis-understanding of the nature of the erosion threat. In several instances, notably the Japanese and Estonia literature, the assumption is made that erosion would result in the removal of beaches and that the complete loss of a precious recreational or cultural resource was imminent. But erosion simply changes the location of beaches; it doesn't threaten their absolute existence. Erosion and beach loss are two separate issues.

A second, much more common and perhaps more important mis-conception about the Bruun Rule, found frequently in coastal management documents, is that it predicts the total erosion. On the contrary, the rule is supposed to predict only the erosion resulting from sea-level rise, a component of erosion that probably ranges from very important to insignificant on various shorelines.

The Bruun Rule was also used by a group of scientists to estimate how much shoreline retreat might occur on the shoreline north of the mouth of the Columbia River in the event of a catastrophic earthquake. But even if the rule did work, it should be applied only to gradual sea-level changes, not instantaneous ones.

New Zealand appears to have provided a particularly comfortable home for the Bruun Rule. Engineering consultant Jeremy Gibb and Peter Cowell, in the School of Geosciences of the University of Sydney, Australia, seem to be leading a national charge to determine hazard zones by adding the Bruun Rule retreat to a fifty- or hundred-year retreat estimate based on present-day erosion rates. An example of this approach is in Hokitika, New Zealand, where a new shorefront development is proposed. This is a rocky shoreline with sand and gravel beaches subjected to fairly frequent earthquake activity. In 1930 the coast was raised a full meter by an earthquake, causing the shoreline to immediately move seaward. Independent of the fact that the Bruun Rule itself is invalid, there could not be a worse place to apply the rule, which requires a tectonically stable, sandy, and rock-free shoreline. Hokitika is none of the above.

In a 2006 legal case (Rob Young, American coastal geologist mentioned in chapter 6, was a part of the proceedings), a judge requested that three coastal scientists separately and independently use the Bruun Rule to predict shoreline movement resulting from sea-level rise over the next 100 years. The location chosen for this test of the rule was a Hawkes Bay, New Zealand, shoreline. The predicted 100-year shoreline retreat numbers were 10 feet, 35 feet, and 190 feet, the variability due to different choices for the closure depth. The point was clearly made. The judge threw out the Bruun Rule in Hokitika.

Figure 5.5 A cemetery falling in along the eroding shoreline of Majro Atoll in the Marshall Islands. Atolls are the "canaries in the mine" when it comes to sea-level rise. They are highly vulnerable to sea-level changes because of their low elevation. Already some, such as the Carteret and Tuvalu atolls, are in the process of being abandoned in favor of higher ground.

Use of the Bruun Rule to determine future erosion rates on the Pacific atolls (figure 5.5) is another particularly lamentable misuse. The aforementioned Peter Cowell and Paul Kench, both senior lecturers at the University of Sydney, seem to be leading this effort. Atoll nations are in serious danger of extermination by sea-level rise. Some, such as the Beaufort Atolls near New Guinea, have been abandoned already. There is no room for error, and using the invalid rule on a rocky shoreface that fails to meet the criteria for rule application is senseless and shameful. Time's a-wastin' for the Pacific atolls, and the planning process for a rising sea level needs a firmer foundation.

Alternatives to the Rule

There are some alternative, adaptive approaches to the Bruun Rule. The first is *extrapolation of present-day erosion rates into the future*. If the shoreline is eroding at five feet per year now, assume that it will continue to do so in the future. This is a widely used approach that takes into account all causes of shoreline retreat, including sea-level rise. It also takes into account differences in the relative importance of various beach processes along different shoreline reaches. But rates of retreat often vary significantly from decade to decade on the same beach. So who knows where the future lies, especially if sea-level rise accelerates as the greenhouse effect marches on.

Another approach is *simple inundation calculations* assuming that coastal land will flood rather than erode. One assumes that after a three-foot sea-level rise, the shoreline would move inland to what was once the three-foot elevation contour on maps. This approach might actually work for quiet lagoonal waters behind barrier islands on low-lying coastal plains.

A third possibility is *a conceptual model of future shoreline retreat* obtained from estimates of future sediment budgets and judgments about the dispersal of beach sediment in the future. This would include qualitative predictions about future storm size and frequency, river discharge, dam construction, seawall and groin construction, beach nourishment, channel dredging, and a lot of other things, including the future behavior of people along the shore. A great deal of educated guesswork and intuition would be needed here, but the result would still be much closer to reality than the result obtained by applying the Bruun Rule.

It is very likely, however, that the Bruun Rule will continue to be applied for years to come. There is no better example of the difficulty of stopping a mathematical model juggernaut. The model's longevity is fueled by the fact that the question being answered is a critical one for society and that no other model claims to solve the problem. In addition, the model is extremely simple, requiring only a navigation chart. Those who use it can take comfort in the fact that they are in the company of many other users all over the world. A few leading scientists and engineers still hawk the model, undaunted by a wall of strong, almost unanimous but quiet protest from the scientific community at large.

In the face of uncertainty we must, of course, make a judgment, even if only a tentative and temporary one. Making a judgment means we create a mental model or an expected universe.

—*Charles Perrow,* Normal Accidents *(1999)*

chapter six

beaches in an expected universe

A Historic Place

Delaware's North Shores is a small, Atlantic Ocean–facing town just south of Cape Henlopen State Park at the entrance to Delaware Bay. Like many other American coastal towns, North Shores has a growing row of expensive houses constructed right next to an eroding beach. In line with Delaware state law, the beach is private, and nonresidents, while allowed to walk on it, must keep moving along. No stopping for a picnic or a swim. Even so, the beach at the engineering groin structure that marks the town's boundary with the state park is listed on gay beach Web sites as an important lesbian hangout.

Clearly visible from the North Shores large stone groin are the most salient features of the park's beach, two World War II watchtowers once used to monitor U-boat activity. The shoreline retreated past them, and both towers now rest in the surf zone. The main reason the park's shoreline has retreated farther inland than the town's shoreline is the groin (locally called a jetty), built perpendicular to the beach in the 1970s. Most beach sand transport here is from south to north, and sand has piled up in front of the town to the south of the groin. Simultaneously the beach on the park to the north has been starved of sand and has eroded. The groin at the north end of the line of buildings has worked all too well.

Yet property owners want still more beach sand to be trapped in front of their property, so they proposed to make the groin higher as well as to install concrete in the cavities between boulders in the structure, to prevent sand leakage through the rocks.

Delaware's state regulators are not supposed to approve projects that will damage adjacent beaches, especially in state parks. So the case seemed to be a slam dunk for those who opposed the permit for the changes to the groin. Several prominent geologists told the permitting agency that the new groin would enhance the erosion problems on the adjacent state park shoreline. One of the geologists, John (Chris) Craft, a retired professor from the University of Delaware, had spent his professional lifetime studying the Delaware shoreline. If the beach were to widen and increase in height next to the new groin, as the town expected it to do (and as all experts agreed that it would), it was clear that it would cut off much of the remaining sand that would flow to the state park.

The small town's engineering consulting firm used a U.S. Army Corps of Engineers quantitative mathematical model (GENESIS) and determined that the higher, tighter groin would not accelerate erosion in the park. It was a virtually impossible conclusion. Yet, unimpressed by the expert testimony, Tony Pratt, the state's coastal manager, and the Delaware Department of Natural Resources and Environmental Protection (DNREP) declared that the consulting engineers were correct. A permit was issued for the groin, and the project was given the go-ahead.

In actuality it was not even clear whether the engineering consultants had really gathered all the appropriate data and observations to make a model run. They provided model results that asserted that the groin would not damage the adjacent beach, but they gave few details of the assumptions they made or explanations of how they came up with their real-world data for the model.

It was a good example of how quantitative mathematical modeling answers, with their alleged state-of-the-art approach, can trump experience, brush away big-name expert testimony, and even defy common sense. It was also a case in point concerning the inability of both policymakers and laypeople to challenge modeled predictions. Tucked into a model, the absurd conclusion that the improved groin would do little harm to adjacent beaches in the state park was acceptable.

Rob Young, a prominent coastal geologist and a professor at Western Carolina University, appeared in 2004 before the Delaware Environmental Appeals Board on behalf of the Sierra Club. But rather than do the obvious and add his voice to those of other scientists who insisted that the groin project would cut off sand to the park, Young chose to attack the mathematical model. He used a three-pronged critique of the case made by the engineering firm.

It must have been a great temptation to Young to simply list the flaws in the GENESIS model (which he and others had discussed in detail in a 1997 article in the *Journal of Coastal Research*). But he was careful to keep his arguments simple and straightforward, easily understandable for a non-technical board. First, Young pointed out that not enough information was given to know whether the model had been applied properly. For example, what was used for wave data? Was an annual mean number used for wave height? Was it the highest one-third of all waves? From what directions did the waves come? What was the method by which wave characteristics were obtained? How do you characterize something as complex as a surf zone for a model? How were storms taken into account? In the GENESIS model it is required that the permeability of the groin (its ability to let sand flow through it) must be known, but that number was not visible, and neither was there a mention of the method used to obtain it (an almost impossible parameter to come by).

For his second point, Young noted that a major assumption in GENESIS is that the beach and nearshore zone, extending to a depth of thirty to forty feet, is entirely covered by sand. This assumption was wrong—spectacularly wrong. Outcropping on the lower beach and exposed at low tide were layers of highly compacted mud containing tree stumps from a forest that lived behind a dune ridge far seaward of the present one (probably 1,000 to 2,000 years old). Clearly GENESIS didn't apply to a stump-and-mud-covered beach and shoreface.

Young's third avenue of assault involved the engineering firm's refutation of the experts. The consultants said that the expert testimony opposing the groin showed geologic bias and should therefore be ignored. This, of course, was akin to claiming that the testimony of a physician in a medical case showed medical bias. Young pulled out the Corps of Engineers' latest version of the *Shore Protection Manual*, the so-called bible of coastal engineering, and read from it. The manual said that all groins cause erosion and that if they are put in place they should be monitored to see what damage occurs and funds for mitigating the damage should be available and in place. It turned out that engineering bias about groins was the same as geologic bias!

After Young's testimony, in February of 2003, the case for the groin virtually collapsed. It was clear that the state had abdicated its responsibility in accepting the model results without examining the details of the model and its basis in the real world. It was the first time, to our knowledge, that a beach engineering project on a North American coast was judged solely because of the fallacies of the quantitative mathematical model that supported a project.

But wait. The law stepped in and the case was thrown out on a technicality having nothing to do with beaches, groins, erosion, or models. As the Sierra Club proceeded to appeal the decision at a higher level, the community threw in the towel and chose to spend $1 million to nourish the beach.

North Shores is a historic place.

Engineers and Beaches

The beautiful beaches where millions come to commune with nature are becoming instead intensely engineered bodies of sand. As people inexplicably crowd buildings up against a shore that is steadily moving toward them, often at a known and well-publicized rate, their homes and high rises soon need to be saved. To the rescue come the coastal engineers, those who practice a branch of civil engineering in which mathematical modeling is the way to go.

The heritage of mathematical modeling followed the engineers to the beach. Modeling concrete and steel for bridges, dams, and elevated water towers is relatively easy. There are few surprises and the designs incorporate large safety factors, so failures are few. Modeling beaches, on the other hand, is very different. They are complex systems that operate under the control of many variables, which are often poorly understood. The various parameters involved in creating beach change work simultaneously and in unpredictable order, timing, and magnitude. There are many surprises. No one knows when the next storm will happen by, and this fact alone wreaks havoc on the neat and orderly world of mathematics at the shore.

No better example of the random storm problem exists than at Sinnes, Portugal, where a few decades back, the government constructed a huge and costly artificial harbor. Almost immediately after its completion, a storm said to be at a 200-year-recurrence-interval magnitude came from the southwest, an unheard-of direction for major storms along this coast. Two years later a *second* 200-year storm roared in from the same direction. The new harbor lay in ruins. The engineers duly noted the rarity of the storms as a reason for their failure, and presumably they were forgiven.

Engineers remedy the shoreline erosion problem in three ways:

- Stand fast with hard shoreline stabilization (seawalls, groins, offshore breakwaters, jetties);
- Stand fast with soft shoreline stabilization (beach replenishment);
- Retreat (move or demolish buildings or allow them to fall into the sea).

If saving the beach for tourists, turtles, birds, fishers, and future generations is paramount, then retreat is the best option. If the highest priority is to save the buildings, then the shoreline must be held in place or *stabilized* with something hard and fixed, like a seawall. A steep environmental price is always paid for hard structures: the degradation and eventual complete loss of the beach. Beach loss occurs as the retreating (eroding) beach backs up against the wall (figure 6.1). Before beach replenishment in the 1990s, the New Jersey communities of Cape May, Asbury Park, Monmouth Beach, and Seabright, all seawalled for more than a century, experienced decades without a beach.

Engineers have been nothing if not ingenious in designing devices to halt erosion. One approach used on Nantucket Island, Massachusetts, has been to pump water out of a beach to reduce the pushing-out effect (pore pressure) of groundwater as it flows from the beach to the sea. Some believe the natural outward pressure of water on sand grains makes it easier for waves to dislodge grains and thus enhances beach erosion. When water is pumped out at depth within the beach, the flow of water at the surface is into and not out of the beach, which tends to hold sand grains in place. Artificial plastic seaweed has been touted as a way to baffle (reduce) waves and currents and cause sand to be deposited. This was unsuccessfully used in front of the Cape Hatteras Lighthouse before the retreat option was chosen in 1999 and the lighthouse was moved back 2,000 feet.

The naming of these erosion-halting devices (sta-beach, sand grabber, undercurrent stabilizers, seascape, wave buster) reflects an equal amount of creativity. The problem is that these approaches almost always

Figure 6.1 Most of Waikiki Beach is beachless, the sand having long since disappeared because of the construction of seawalls such as those in the foreground. This seawall effect is the reason that beach nourishment, with its requirement for mathematically modeled predicted life spans, has come to national prominence.

create more problems than they solve. If they cause sand to accumulate at one point on the beach, which is usually the desired end result, they rob sand from another point. Something like robbing Peter to pay Paul. But beaches must be free to survive and thrive.

Between the extremes of retreat and seawallling, there is a third path—soft stabilization or beach nourishment, which is sometimes called beach replenishment or, most realistically, dredge and fill. This involves bringing new sand to the beach. Hundreds of U.S. beaches from

Figure 6.2 Beach nourishment on Emerald Isle, North Carolina, in 2005. This sand was being dredged and pumped to the beach from the tidal delta at a nearby inlet and then leveled on the beach with bulldozers. Photo furnished by Rudi Rudolph.

Figure 6.3 Another nourished beach on Emerald Isle, North Carolina (2003), made up mostly of broken oyster shells. This material is so coarse that shoes must be worn on the beach. Replacing a sand beach with such shell gravel does serious damage to the nearshore ecosystem of a sandy beach. Unsatisfactory material on this beach resulted from poor pre-nourishment surveys of the continental shelf offshore from the beach. Photo by Andy Coburn.

Waikiki, Hawaii, to Galveston, Texas, to Miami Beach, Florida, to Atlantic City, New Jersey, have been repeatedly replenished—some more than twenty times since the early 1960s.

The new beach sand comes from a variety of sources, most of it pumped from the continental shelf by dredges. Some communities—for example, Virginia Beach, Virginia—bring sand in by dump truck. Waikiki Beach once obtained Southern California sand that had been transported to Hawaii by freighters. Beaches on all coasts sometimes receive sand obtained from routine navigation channel dredging. The whole process is very costly, $1 million per mile of beach at a minimum, and sometimes much more. Most of it has been paid for by the federal government. A few years ago, it cost in excess of $200 million to pump up a large artificial beach along twenty-one miles of the northern New Jersey shoreline. That's $10 million a mile.

The most widespread use of the mathematical modeling of beaches today is for the design of nourished beaches (figure 6.2). There are more than 225 such beaches on the U.S. East Coast and Gulf Coast barrier island shorelines. In order to receive federal funding, each nourished beach project must achieve a favorable cost-benefit ratio. This means that the life span of the artificial beach must be predicted in order for the cost to be figured. What will the rate of sand loss be and how soon will the beach have to be replaced? If only we could just dump sand on the beach, smooth it out, and walk away, ignoring for the moment the vast impact that artificial beaches have on plants and animals of the near-shore (figure 6.3)!

If it weren't for the cost-benefit ratio problem, beach modeling would probably reside only in academia and not be a part of our society's political maelstrom. Ironically, the Dutch, whose Delft laboratories have probably devised more mathematical models of beach behavior than any other group in the world, do not require accurate prediction of the beach life span before nourishing a beach. Once the Dutch decide to nourish a beach, they just "dump and run." It makes life much simpler.

The River of Sand

Most beaches can be described as rivers of sand. Just as in a river, large quantities of sand are moved within a narrow band by currents. But instead of flowing downhill under the force of gravity as rivers do, the force of breaking waves approaching the beach at an angle creates the current that carries the sand on a beach. Almost always, waves come in at a slight angle to the shoreline, and as they break, some of their momentum is transported laterally in one direction or another, parallel to the beach. For example, on a north-south-trending beach, waves from the south will move sand to the north and waves from the north will move sand to the south. Sand flowing in two directions at different times is another important difference between a beach and a river.

How much sand is moved by these wave-formed surf zone currents is a critical question. The gross sand volume is the total amount moved in both directions. Subtracting the smaller from the larger gives the *net longshore sand transport volume*, a number often used in mathematical models of beaches. On the Outer Banks of North Carolina, summer waves carry sand to the north. The direction reverses in the winter, and since winter waves are larger than summer waves, net longshore transport of beach sand is to the south. On the beach, the "downstream" direction (the direction of net sand transport) is called *downdrift* and "upstream" is *updrift*.

Other things being equal, the bigger the waves and the stronger the current, the more sand is transported. An integral part of the transport process is suspension of sand by the breaking waves, a phenomenon well known to any and all who have ever waded into an ocean surf zone. Because breaking waves toss sand into the water column, even very small longshore currents, at current velocities too slow to pick up individual sand grains from the sea bottom, can still carry a lot of sand.

The volume of beach sand transport can be huge. At Santa Barbara, California, perhaps a million cubic yards, plus or minus, per year (net sand transport) move from north to south. That's around 100,000 dump truck loads. Along the aforementioned southeast coast of Iceland, one of the highest-wave coasts in the world, an order of magnitude estimate indicates that, in some years, as much as 5 million cubic yards of sand may move from east to west annually. On Bogue Banks, North Carolina, field indications are that net transport is close to zero. It all depends on the nature of the waves striking a beach, where they come from, how big they are, and how much sand is available to be moved. (Both the Santa Barbara and the Icelandic sand volumes mentioned above are rough estimates at best and probably vary considerably from year to year.)

It's difficult to measure sand volumes in the surf zone. In fact, accurate characterization on even a short-term basis has proved to be impossible. One way to get at short-term transport (hours) is to use tracer sand grains dyed with some fluorescent color. In the 1970s radioactive tracers were used, but although they were superior to fluorescent grains in ease of detection, public fears about the health effects of radiation quickly halted those experiments. Sand traps and electronic devices that detect suspended material in the water column are other measurement approaches.

An indirect, longer-term approach is to measure the changes in the shape of beaches and other sand bodies, such as tidal deltas, at inlets. Losses or gains can be an indication of longshore transported sand. Other methods take advantage of coastal engineering and measure how quickly a dredged channel fills up or how much sand gathers in the lee of a jetty next to an inlet.

All such indirect measurements really address how much sand was accumulated, not how much sand was transported. There can be a large difference between these two quantities. Furthermore, there is no guarantee where such sand came from or how it got there.

The biggest problem of all is the measurement of sand transport in storms. This is when the most sand is transported, when the most changes occur on beaches, and when engineers and geologists are forced to pick up their instruments and beat a hasty retreat. In the really big storms, the surf zone may become hundreds of yards or even miles wide, and sand may be transported in a band that is at least as wide as the band of breaking waves.

Sam Smith, an Australian coastal engineer, thought he had a solution to measuring the storm problem. He spread a dump truck load of blue-dyed fluorescent sand along a Gold Coast beach in Australia as a typhoon approached the shore. Smith figured he would be able to find at least a few grains and thus learn the direction in which sand was moved. After the storm, however, not a single blue sand grain was to be found.

Waves: The Driving Force

Waves seem to have the same hypnotic effect as campfires. Standing on a beach, we are all transfixed by them. They arrive with a roar and end with a whimper as a thin sheet of water quietly bubbling up the beach. Their variety is endless. Large and small, they all break on the beach and die there. The larger waves break well offshore and transform themselves into smaller, more frequent waves that break again closer to the beach.

The pattern of breaking waves depends to a large degree on the shape of the beach. For example, offshore sandbars trip the waves and cause them to break offshore. One or more distinct lines of breaking waves mark the location of offshore bars. Among natural catastrophic events, a surf zone in a giant storm ranks right up there with volcanoes, landslides, and tornadoes for awe-inspiring vistas.

Surfers probably understand waves better than any other group of beach dwellers, with the possible exception of those lifeguards who have a sense of curiosity about nature. Surfers spend hours swimming through, sitting in, and riding down the waves. They prefer the widely spaced (long wavelength), regular waves formed by winds a long way out at sea, which oceanographers refer to as *swell*. Local storms produce more irregular waves, called *sea*, a "confused" surface of more closely spaced and less uniform waves.

Southern California surfers habitually watch the Pacific-wide weather forecasts, knowing that winter storms in the Southern Ocean—for example, off New Zealand—can produce a fine surfing swell in California. During the winter, storms off the Aleutian Islands may produce the best surfing of all. The surfers also know that because of differences in beach orientation and beach shape, some beaches are best for the New Zealand waves and others are best for the Aleutian waves. Sometimes, however, local winds may mess up the beautiful swell from distant oceans. Waves will come from two or more directions, creating a chaotic surf zone usable by only the most die-hard surfer, who must be willing to settle for short rides.

Those who would apply math to waves must do so by simplifying the surf zone, including both the incoming waves and the seafloor surface on which the waves break. It is hard to imagine a greater contrast than that between the pandemonium of waves in a confused surf zone and the disciplined regularity of mathematically described and averaged-out waves.

The single most important characteristic of waves in the mathematical models is *wave height*, the vertical distance between the trough and the crest of a wave. Beach modelers have described wave height as the "Rosetta stone" of models. The higher the waves, the more the sand moved and the greater the beach change. More than one nourished beach has been quickly lost to the high waves in a single storm. The high waves of a 1971 storm at Cape Hatteras, North Carolina, removed an entire beach overnight, just days before its restoration by sand pumping was completed.

The idea in the design of artificial beach is to determine wave heights in the past and project them into the future. The assumption is made that the waves in the future will be just like the waves in the past. Normally, average wave heights are determined in time increments like weeks or months. Sand transport is then determined using a single wave height for each increment, and then all the time spans are added together to come up with annual net transport volumes, the unit of measure used in most beach engineering projects.

So how do engineers come up with wave height on a particular beach that can be used to describe the amount of sand moved on a beach over months and years? It is usually a long and tedious five-step process, made necessary because real-world, long-term surf zone wave data essentially never exist where an engineering project such as an artificial beach is proposed.

Step 1. Predict deepwater wave height and direction. Deep water begins at the depth where the incoming waves just begin to stir up the bottom. The wave height is determined by a hindcast, i.e., consulting a table of wave heights calculated from past weather conditions. The relationship between hindcast waves and real waves remains largely unknown. The requisite studies to test the accuracy of these theoretical waves have not been carried out.

Occasionally, actual offshore wave measurements from a so-called wave buoy anchored beyond the surf zone will be available. Wave direction can be difficult to determine because waves generated by different winds in different patches of the ocean may arrive at the beach simultaneously from several directions at once, a phenomenon frequently visible from fishing piers or from the air. Which direction of wave approach do you use? To "solve" this problem, engineers assume that at any given point in time, all waves come from the same direction.

Step 2. Bring the waves ashore (on paper) from deep water into shallow water. This is accomplished using a mathematical model, which, given the deepwater wave height and direction, predicts how the waves will bend (refract) when they come ashore. The problem is that nearshore topography, which often changes over time, must be well known (figure 6.4). Either real topography is used or a theoretical topography is used, based on a so-called profile of equilibrium. In the latter case, the shape of the nearshore surface (the shoreface) is determined, using a simple mathematical equation based entirely upon grain size of the sediment covering the shoreface. As the waves roll ashore, friction between waves and the seafloor affects the height of the waves in the surf zone, Smooth rock surfaces on the seafloor absorb the least wave energy, while medium sand-size sediment absorbs the most wave energy. This friction effect is not considered.

As a side note, careful (and skilled) observers can sense the presence of offshore rock outcrops by staring at the surf zone and noting

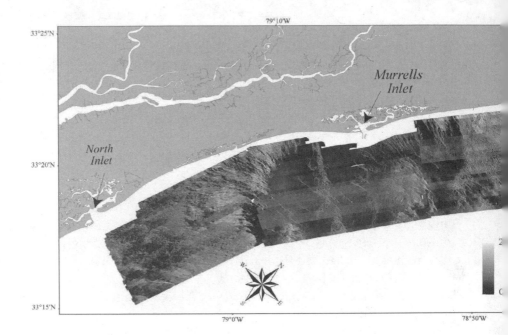

Figure 6.4 Side-scan sonar mosaic of the seafloor off Myrtle beach, South Carolina. This is a "photograph" made by sound and showing the rocky and patchy nature of the shoreface. Models predicting the behavior of beaches assume that the shoreface is entirely covered by sand of uniform grain size, which is virtually never the case. Photo furnished by Jane Denny and the U.S. Geological Survey.

variations in the patterns of the waves. A few years back, standing on a fishing pier and observing wave patterns in the surf zone in Wrightsville Beach, North Carolina, Sam Smith, the aforementioned Australian coastal engineer, pointed to the location of an offshore rocky area, which, a year later we "discovered" by using side scan sonar.

Step 3. Choose a wave height. There is no better example of the rigidity of models, their lack of flexibility, and their lack of reality than the choice of a single average value to describe the confusion and jumble that is the surf zone. Reasoning that higher waves transport more sand than smaller ones, the highest one-third of the waves over a given time frame (weeks to a year) is often chosen as the breaking wave height to be inserted into models. In some models, those concerned with storms, the average of the highest twelve-hour wave height over a time frame as long as a year may be used. But who knows what measure of wave height will apply to a particular beach?

There is no field evidence to support the choice of any single measure of wave height for use in models; if such a number did exist it would

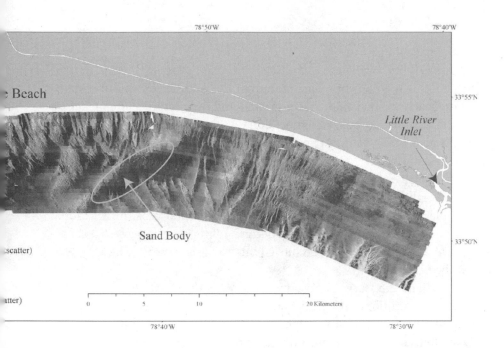

Beach

Little River Inlet

33°55'N

Sand Body

:scatter)

33°50'N

atter)

0   5   10   20 Kilometers

78°40'W   78°30'W

surely be different on different beaches. Averaging parameters in models removes the sharp corners and the all-important extreme events. In a study of Myrtle Beach, South Carolina, for a beach nourishment project, for instance, a consultant averaged the waves seasonally, or on a three-month basis. Gone (smoothed out) were the storm wave heights, the very events that move the most beach sand on this particular beach.

Step 4. Break the waves. The shape of the shoreface determines how waves will break, how they will feel and interact with the bottom, and to a large extent how much sand they will transport. An unchanging "design beach profile" is assumed, usually without sandbars or other complications. In reality, sandbars often control sand transport, and their impact constantly changes as waves move sand and change the shape of the bar or as the tide goes up and down or as storms come and go.

Step 5. Move the sand. In most models this is done using an equation (like the CERC equation described below) that relates breaking wave height to sand transport. Actual accurate measurements of net sand transport in the surf zone have proved impossible to obtain, so this remains guesswork.

Each step in this process of determining the wave height, the single most important parameter in mathematical modeling of beaches, is fraught with error and unreality. Three or more quantitative mathematical models are used in step-like fashion to get the requisite wave height number. The Rosetta stone of modeling is made of clay.

Mathematical Beaches

Since direct methods of measurement of the volumes of sand transported on beaches (such as by tracer grains) and indirect measures (such as sand accumulation behind jetties) don't work, mathematics must surely be the answer. The U.S. Army Corps of Engineers, keeper of the nation's beaches, devised the Coastal Engineering Research Center (CERC) formula to solve the problem (see appendix). The CERC equation uses wave height, sand grain size, and angle of wave approach as the main variables. It has been the mainstay of American and most international beach studies for more than three decades. But since no way exists to accurately measure what's been moved, no way exists to know how close to reality the answers are.

Back in the early 1970s, two scientists, Douglas Inman, of the Scripps Institution, and Paul Komar, then a graduate student, tried to test the equation. They used fluorescent tracers on two California beaches in fourteen two- to four-hour experiments to measure how much sand was moved. Simultaneously they measured the waves to provide the basis for calculating the longshore transport by a CERC-like formula. Of course the field and calculated values didn't match, so a K, known in the parlance of coastal engineering as a *sediment transport coefficient*, was inserted into the equation to make the two numbers the same. The coefficient K was said to equal 0.77, and it was multiplied by the model's derived or calculated longshore transport volume number to come up with the correct result.

To accept 0.77 as K, one must also accept as valid the longshore sediment transport volume determined in the field by tracer sand grains. In addition, it must also be assumed that the volume determined by the CERC equation is proportionately correct as well but is consistently slightly too large and can be corrected by multiplying by 0.77.

Neither assumption is valid. Sand tracer studies under non-storm conditions provide only poor estimates of longshore transport volumes, and the CERC mathematical model is too simple, failing to consider many important factors that may move sand. Multiplying K, a coefficient determined by a poor field technique, by an incorrect volume determined by the weak CERC equation cannot provide a correct sand transport volume. There simply is no basis to assume that K, the sediment transport coefficient, is valid.

The fallacious K has persisted, however, and sand transport on many U.S. and European beaches has been and continues to be calculated using the 0.77 determined in brief experiments on a calm Southern California beach to "correct" the equation. Inman and Komar allegedly "solved" an important problem, and no one has looked back. No one lifted the flap of the tent and looked underneath at the absurd basic assumptions.

In 1998, Ping Wang and Nicholas Kraus measured some East Coast and Gulf Coast beaches using three- to five-minute measurements of suspended sediment quantities in sand traps placed within the surf zone. They concluded that K should be 0.08 (for moderate wave energy beaches). Imagine using a three- to five-minute measurement on a beach to determine a coefficient that determines the annual longshore transport, storms and all. As in the case of Inman and Komar's work, it simply defies all logic. And the difference between the 0.77 and 0.08 multiples is an order of magnitude. In fact, coefficients as high as 1.15 have been determined, expanding the range of K values to a range of two orders of magnitude.

Use of the aforementioned U.S. Army Corps of Engineers mathematical model called GENESIS (see appendix) requires the use of two K's. The model was used in a 1999 design plan to figure out how much sand would be transported along the beach in the vicinity of proposed jetties at Oregon Inlet, North Carolina. The report estimated that the annual volume of net longshore transport of sand was in the vicinity of half a million cubic yards to the south, approximately the same as two earlier studies (each using a different figure for K). The corps argued that the similarity in sand transport volumes of the three separate studies indicated that the quantitative model GENESIS must have been working. But the K values in the most recent study were specifically chosen to come up with the same number as the previous studies, so of course the numbers matched.

In another Corps of Engineers study on North Carolina's Outer Banks, the volume of sand that was expected to flow to the south from a beach nourishment project at Nags Head, North Carolina, was calculated. The model-projected sand transport volume was so large that there was a danger that sand would clog up the navigation channel at nearby downdrift Oregon Inlet. The increased channel dredging costs at the inlet would make the nourishment project all the more expensive, and worse yet, the required cost-benefit ratio would be unfavorable.

The solution? Go back to the mathematical models and change the predicted angle of wave approach to the shoreline to reduce the theoretical southerly sand transport a bit. The problem was solved! Later the engineer (from the Corps of Engineers' Vicksburg, Mississippi, research center) who accomplished this money-saving feat explained it in detail to a group of disbelieving federal agency representatives who had met to discuss the project. He noted, with some pride, that by merely changing the wave angle he had removed an obstacle that could have blocked an important nourishment project (projected to cost $1.6 billion over the next fifty years for nourishment along fourteen miles of shoreline). Mathematical models made possible his utter detachment from reality, and the agency representatives who listened to his presentation learned the hard facts of life about politicized models.

This kind of customer-driven modeling could well be defined as brazen intellectual dishonesty. Some would (sarcastically) describe the process as objective analysis—you provide the objective, and I supply the analysis. So it went in the North Carolina Outer Banks beach nourishment mathematical modeling. The local Corps of Engineers district office stated that the objective was less sand flowing to the south, and the Vicksburg modelers made the appropriate analysis to achieve the desired objective.

Above and beyond the K problem, the idea that a single model such as the CERC equation can be used on all beaches defies the principle widely accepted by coastal scientists that all beaches are different. The CERC equation adheres to historian Arnold Toynbee's observation that "the price of quantification is the loss of uniqueness."

The huge number of variables that control beach behavior, variables that vary from region to region and from beach to beach, illustrates what complexity is all about. There is no better indication of the real-world meaning of complexity than a list of parameters that affect longshore sediment transport. Myriad different processes can be involved, each to different degrees of importance on different beaches or on the same beach at different times. This is the ordering complexity phenomenon—an aspect of complexity mentioned briefly in chapter 5, where a similar array of parameters is listed that affect shoreline retreat. The point is that prediction of the future is impossible, because even if all parameters are understood, the order in which they will take effect can never be known.

The following subjective listing of longshore transport parameters is arranged in three categories of relative global importance, with asterisks denoting the eight parameters explicitly and implicitly considered in the CERC equation (see appendix). Clearly the CERC equation (and a whole fleet of other models slightly modified from the original CERC equation) only scratch the surface of the world of longshore currents.

Always Important
*wave height
*wave period (time for passage of two wave crests)
*angle of wave approach to the shore
storms
*shoreface shape
feedback (continuous offshore bar shape changes)
*grain size
underlying geology (rock and mud layers beneath beach)
*water depth

Usually Important
offshore bar shape and location
interactions of waves and currents
wave setup
wave energy loss by friction (with seafloor)
seaward sand transport by waves
seaward sand transport by currents
loss or gain of sand from wind transport
loss of sand from storm overwash
coastal type (e.g., rocky, sandy, reef)
sediment supply
engineering structures (e.g., revetments, breakwaters)
beach nourishment
beach rock
nearshore winds
shell pavements
bedforms (e.g., ripple marks)
bottom roughness

Sometimes Important
bed liquefaction (quicksand formed as wave impacts)
beach state (e.g., pre- or post-storm)
storm surges
tidal range
tidal currents
offshore coastal currents
sea water infiltration
wave types
*wave-breaking parameters

wave reflection
infragravity waves
wave reformation (after breaking)
water temperature
sediment sorting
beach stratigraphy (vertical layering)
shape of gravel and larger clasts
*specific gravity of sand grains
pore pressure from groundwater flowing from beach
organic mats on beach surface
downhill flow of dense, sediment-filled water
sediment churning/burrowing by organisms
rip currents through the surf zone
seaward return of storm surges (storm surge ebb)

Nine parameters in the list are in the Always Important category, seventeen are considered Usually Important, and twenty-three are in the Sometimes Important category. The categorization of these factors is purely arbitrary; any of the forty-nine parameters could be of overriding importance in particular circumstances.

An example is nearshore winds, in the Usually Important category of the list. This is a parameter that is not employed in any of the beach mathematical models. During a storm, if the wind is blowing parallel to the beach and in the same direction as the longshore currents formed by waves, the water movement is reinforced and intensified, and the result can be awesome. During a January 1, 1987, nor'easter on Topsail Island, North Carolina, we observed that the combined 30-degree wave angle of approach and strong 40-mile-per-hour winds blowing to the south (parallel to the beach) produced a surf zone that resembled a mountain stream in flood. The intense boiling water, discolored and opaque from its high sand content, rushed to the south. Bouncing and tumbling along with the current were dozens of wooden dune walkovers that had been yanked from their foundations up the beach. This particular storm must have had a big impact on the net volume of sediment transported to the south in 1987, but the predictive models would not have considered it.

Wind also can affect the character of the breaking wave, which in turn affects volumes of sand transport. Any veteran surfer will tell you that an offshore wind will help hold a wave face up, while an onshore wind will "close up" the wave.

The importance of this (incomplete) list of factors that affect the longshore transport of sand is that it illustrates the impossibility of accurate prediction. One can conjure up many combinations of processes acting on a particular beach at a particular time, which is what makes this a complex system. However, to determine only the direction of sand transport, a question for a qualitative model, only the Always Important factors need be considered, as a rule.

When asked about the uncertainties facing the U.S. troops who were about to invade Iraq in 2003, Pentagon planners said that there were categories of uncertainties: *knowns, known unknowns,* and *unknown unknowns.* For example, a "known" was the location and type of bridges crossing the Tigris River, a "known unknown" was the will of the enemy to stand and fight, and an "unknown unknown" was the breakdown of the Iraqi society when Baghdad fell.

The list of parameters that affect longshore transport can be viewed as uncertainties just like the parameters of a war campaign in a distant land. But in modeling sand transport on beaches there are no knowns like a Tigris River bridge because even the Always Important parameters, like wave height and shoreface shape, are imperfectly known. All the parameters in the table are known unknowns. The unknown unknowns on the beach are the processes and combinations of processes on the seafloor and in the surf zone that are yet to be discovered. A newly discovered process (found by Donn Wright and his associates at the Virginia Institute of Marine Sciences) is the recognition of a previously unknown type of seaward water flow on the seafloor, driven by gravity alone. During storms, this flow is capable of moving fine sand, and especially silt, for long distances down very gentle seafloor slopes of a fraction of a degree. As the flow moves seaward, waves keep the sand and silt grains suspended in the flow. The problem of modeling the transport of sand on beaches can be summed up as a "cascade of uncertainties." It shows the quadruple interaction of air, sea, sediment, and people, beginning with the imperfectly understood components that influence sediment movement (climate and oceanographic processes), proceeding through the events in a surf zone to the vagaries of engineering practice, and ending with the unpredictable behavior of humans who decide on coastal management policy and play politics with sand volumes. A successful mathematical model must wend its way through all of these uncertainties to come up with a "correct" answer.

The Cascade of Uncertainties in the Prediction
of Longshore Sand Transport on Beaches

Climate variability.
     Atmospheric processes
         Oceanographic processes
             Longshore transport processes
                Sediment morphology changes
                   Engineering activities
                     Policy
                       Politics

Bad Numbers That Stick Around

Once longshore transport volumes are modeled for a given shoreline reach, they become a form of conventional wisdom and are cast in stone. Seldom do engineers (or scientists, for that matter) look back at the numbers. It is also universally assumed that annual transport volume will not significantly change from year to year. But the fact is that the annual transport volumes *are* likely to be vastly different from year to year.

Annual sand volumes, such as those mentioned above for the Outer Banks, have lasted unchanged in technical publications for decades. Use of the extant values is the simplest way to go, the path of least resistance. We have a number, so let's not complicate things by updating, recalculating, or checking. Let's get down to the task at hand and use that number.

Charles Perrow, retired Yale professor and author of *Normal Accidents*, would probably characterize the volumes of longshore transport on beaches as existing in an *expected universe*. Our lives are governed by innumerable expected universes, defined by our past life experiences. But an expected universe based on an uncertain judgment (such as the CERC equation) is dangerous. The possibility, even the probability, exists that the range of net longshore transport volumes currently accepted as normal (100,000 to 1,000.000 cubic yards per year) and the assumption that each year roughly the same net sand volumes are transported in the same direction are wrong on many beaches. For example, in the 1987 storm that we observed, net longshore transport volume of sand on Topsail Island, North Carolina, must have been much larger than average.

Reality in longshore transport volumes on beaches remains elusive, particularly because of our inability to measure sand transport during storms. Thus we have created an entirely artificial expected universe of longshore transport volumes. When an annual net longshore sand transport number threatens to leave the universe, it is brought back in line by fudge-factor adjustments or the choice of "more reasonable" values and assumptions for model parameters.

It is a truth of human nature that escape from an expected universe is difficult in the extreme. Quantitative mathematical modeling using equations that assume the expected universe exists make escape all the more difficult. Yet escape from the longshore transport universe we must, if future nearshore sediment studies and their engineering applications are to be meaningful.

If the Beach Models Are All That Bad, Why Isn't It Obvious?

Coastal engineering failures, such as the unpredicted rapid loss of an artificial beach, are never as obvious to our society as the collapse of a bridge or the rupture of a dam is. The public has a very short memory about the projections of beach life spans. Often the need for renourishment of a beach comes two or three years or more after the dust has settled from the thorny societal debate that preceded the nourishment project. The players and the issues have faded beyond recall. Local politicians, eager to keep federal funding for future renourishments, are not about to bring up the subject of the failed predictions and an unrealistic cost-benefit ratio by the agency (usually the U.S. Army Corps of Engineers) that paid for the beach.

When artificial beaches are lost more rapidly than predicted by the models, the most common excuse is that the storm that caused the beach loss was unusual and unexpected. Certainly unusual storms can occur, but the label "unusual" is used so frequently with lost artificial beaches as to imply that the last few decades have been truly extraordinary in their storminess. Occasionally the "unusual storm" excuse is even laid out ahead of time. One engineer from the Corps of Engineers stated in 1981 that the new nourished beach on Miami Beach would last forever, unless an unusual storm came by.

A nourished beach emplaced on Ocean City, Maryland in 1991 largely disappeared by January 1992. The cause? Two unusual storms, one in November 1991 and the other in January 1992. They were declared by the city to be 100- to 200-year storm events, which explained why a ten-year beach lasted one year. Making the loss of the beach even more acceptable was the claim that the beach had saved the city $93 million in storm damage. The beach had also salvaged the city's $500 million tourist industry. Quite a feat for a beach that cost a mere $60 million, but the 100-year storm designations were exaggerated, and the other highly questionable numbers were basically pulled out of the air and never documented or explained by the city or the Corps of Engineers.

Another common interpretation of the premature loss of an artificial beach is that the lost sand is just offshore, continuing in its function of storm protection. This explanation was employed in 1992 at Ocean City, Maryland. Needless to say, the argument is unimpressive to local politicians, who can't bring back the tourist trade with an underwater beach.

Criticism of quantitative beach modeling is either avoided or suppressed by applied beach modelers. Instead of debating the issue in order to recognize the limitations of models, consultants and the Corps of Engineers erect a stone wall between the modelers and their critics. In this regard, the difference between the engineering mentality and the scientific mentality looms as a huge factor. Young scientists view it as their solemn duty to knock the revered old-timers off their perches, constructed over decades of pronouncements and discoveries. Young coastal engineers have more of a tendency to revere their elders. Research scientists thrive in an environment of criticism and analysis, but in the technical coastal modeling literature, criticisms are nearly nonexistent and the models remain unchallenged. To criticize is almost an ungentlemanly thing to do.

Over the years I (Orrin Pilkey) have published a number of criticisms of mathematical models used in beach engineering. All of these papers have been in the science literature, mostly in the *Journal of Coastal Research*. The editor of the principal coastal engineering journal of the American Society of Civil Engineers informed a mutual friend that he would never allow a paper by me to be published. And he kept his word.

A reviewer of a paper I wrote condemning beach models penned the following criticism, which is very typical of the responses that model critics receive: "Everyone, even the engineers, realizes that models have shortcomings, some serious ones, but that is all that they have at this time. They are constantly working on improving them. Instead of continuing to tear down the existing ones, the discipline would be much better served by offering better alternatives."

My response (had I been given a chance to respond) would have been this: One should not use bad models for any reason. If you know there are problems, shame on you and your fellow modelers for not saying so when you apply the model and give the results to the public. Because of the complexity of beaches, rest assured that nothing better is coming along. They can never be quantitatively modeled with sufficient accuracy for engineering purposes. Although some improvement in the models may be possible, the engineering profession is not looking back to see what the problems are. Maybe some stiff criticism might jar engineers into a realization of the absurdity of quantitative beach modeling.

Robert Dean, a University of Florida professor, a member of the National Academy of Engineering, and perhaps the nation's most visible coastal engineer, published a textbook in 2002 on the topic of coastal processes from an engineering viewpoint. The book provides much modeling advice but fails to respond to or even acknowledge the criticisms concerning modeling that are in the literature. In Dean's book and in much of the engineering literature, the science of coastal processes is strongly influenced by model simplifications held dear by the engineering profession. For example, the pioneering work on shorefaces done by Donn Wright, of the Virginia Institute of Marine Sciences, and by Rob Thieler, of the U.S. Geological Survey, doesn't penetrate the model-simplified concept of the shoreface in the book. The same can be said for the leading science textbook on coastal geology. Paul Komar's 1998 second-edition textbook, *Beach Processes and Sedimentation*, gets into models extensively but has nary a whisper about published criticisms of them.

Dean, along with James Houston, who was then head of the U.S. Army Corps of Engineering's Coastal Engineering Research Center, once noted (in a review of *The Corps and the Shore*, by Orrin Pilkey and Kathy Dixon) that "[coastal engineering models are valuable] as a learning structure for making predictions prior to construction [of nourished beaches], subsequently monitoring the project, then later comparing predictions with monitoring results." If only this were true. The Dean-Houston view is a common refrain from engineering modelers, but it expresses a reality that doesn't exist. Long-term monitoring is almost nonexistent, and meaningful objective comparison of model predictions and the results is virtually nonexistent.

Paul Komar, the co-inventor of the K coefficient, characterizes model skeptics (with specific reference to the senior author of this book) as neo-Luddites. Many share his view that model critics are out of date. Komar further argues that the maturity of a specialty science like ours can be measured by the success of its numerical mathematical models and that by this measure coastal science is mature because of its excellent beach behavior models.

We believe the opposite is true. The state of modeling of beaches is a reflection of a most backward science. In fishery modeling, those who devised the models express concern about model misuse, especially by managers who have no understanding of model weaknesses. The fallibility or reliability of global change models is widely and openly discussed. Those who invented the beach models, however, express no such concern, brook no criticism from the outside world, enter into no debate about the models, encourage model use under almost any circumstances, and consider skeptics to be unworthy of a response. The coastal engineering profession is a disgrace.

Alternatives to the Models

As in other fields where quantitative mathematical models hold sway, it will be most difficult to turn things toward a more valid, more qualitative approach in this field as well. The first and foremost hurdle will be the resistance of many coastal engineers even to admitting that there are serious shortcomings in their modeled predictions. The large front-line consulting companies that design nourished beaches and seawalls and provide expert testimony in a wide variety of beach legal actions have a vested interest in the status quo. They have convinced the legal system that models are a must. The public is impressed with models on beaches too, largely because the engineering community has long educated them to believe that models are sophisticated, state-of-the-art techniques. Besides that, the models give solid answers and provide a large fig leaf for politicians who are trying to convince voters to vote for another nourished beach.

First, the legal obstacles to a qualitative approach must be overcome—a formidable problem, to say the least. The requirement of a favorable cost-benefit ratio for most federal beach projects is an outgrowth of misguided or overconfident assertions from scientists and engineers of a previous generation. These earlier practitioners of coastal science and engineers must have assured bureaucrats and legislators that accurate predictions of artificial beach life spans were possible. We now know they are not possible; we have a thirty-year record of consistent failure of models to provide accurate predictions, which of course is the basis of the cost estimate for the cost-benefit ratio. The Corps of Engineers, an agency that must have projects if it is to survive, has no particular incentive to change to a qualitative approach. And beachfront communities and coastal zone legislators certainly will resist any change that might damage the flow of federal money to their communities.

Second, the public must be educated in order to counter the claims of the Corps of Engineers and other modelers as to their assertions of accurate projections. The belief is ingrained in the technology-loving American public that quantification is possible for any natural process and that if we can't do it now we will be able to accomplish it with just a bit more research effort. Of course, both of these assumptions are wrong.

Some engineers argue that model we must because there simply is no satisfactory alternative to models. But that is not true. There are useful alternative approaches to predicting beach behavior. For example, in nourished beach design, one could do the following:

- *Use the kamikaze approach,* or dump and run. Put sand on the beach and see what happens. This actually is carried out often on U.S. beaches, especially when sand from navigation channel dredging is used. In this case, there is no legal requirement for the calculation of a cost-benefit ratio.
- *Learn from the experience on neighboring beaches* that have already been nourished. There are strong regional differences in nourished beach stability (beaches in Florida last much longer than beaches in New Jersey), and this approach is probably the best way to predict beach longevity.
- *Try the Dutch approach.* Determine the rate of natural loss of sand on a beach and simply put up enough sand to counter it. The volume of sand predicted would depend on the number of years of beach life span desired.

Coastal engineering is embedded so deeply in the modeling approach that the way out is not clear. How do we come up with cost-benefit ratios for projects if we admit that the cost numbers can only be crudely estimated? Can engineers swallow their pride and ask Congress for money for beach engineering projects while admitting that they cannot accurately project beach life span? Will Congress accept vague estimates of costs? Will the time come when storms that remove beaches become normal and expected rather than unusual and unexpected?

It can only be hoped that the North Shores, Delaware, case is the beginning of the end of blind and unquestioning quantitative mathematical modeling of beaches. It's long overdue.

Take Deep Throat's advice. Follow the money and you'll know the model bias.

—*Robert Moran, hydrogeological consultant*

chapter seven

giant cups of poison

The Berkeley Pit

Seventy to 80 million years ago, about the time the dinosaurs were breathing their last, the huge body of granite now known as the boulder batholith was forming deep within the earth's crust. Seventy miles long by twenty-five miles wide, the mass of once molten magma extends from Helena, Montana, to the highland mountains south of Butte, Montana. When it was still very young, hot mineralizing fluids traveled outward from the molten rock along cracks and fissures into the surrounding solid rock, to deposit a large variety of minerals that contained gold, lead, silver, copper, arsenic, molybdenum, and selenium. Here the metal-bearing minerals remained for tens of millions of years, slowly coming closer and closer to the earth's surface as overlying layers of rock were gradually peeled away by erosion.

Millions of years after the ores that would someday be mined were first formed, earth movements, part of the building process of the Rocky Mountains, created more fractures and faults throughout the deposit, confusing the mining picture. Rich veins containing silver and copper would abruptly end at a fault plane, only to continue tens or hundreds of feet away, where they had been displaced by the fault. It made challenging work for geologic sleuths and also precipitated legal action and fisticuffs between mine owners as they argued over ownership of the veins.

Lewis and Clark passed through the neighborhood in 1805, but there was no way they could have known they were trudging past one of the world's most precious mountains, one that in seventy-five years would be called the "Richest Hill on Earth." The Native Americans knew that something was different about this mountain. In 1856 when the first wandering white men reached the area on the western flank of the Rocky Mountains, they discovered a small pit, four or five feet deep. Scattered around this prehistoric prospect hole were sharpened elk horns that had been used as picks.

In 1864, during the Civil War, the dam burst when gold was discovered in Last Chance Gulch. The gold rush was on, but in a decade the small, low-grade placer deposits in river gravel and hillside alluvium were all but mined out. Mined from hard-rock deposits, silver came next, and was an important part of the Butte economy until it cratered in 1893 when the United States went off the silver standard.

What made Butte different from many other gold rush towns in the West was copper, a whole mountain of it. At first copper was a curiosity, because it gave the rocks a beautiful blue and green sheen. But by the mid 1870s, when the population of Butte was 1,000, a new national demand for copper arose because of power and telephone industry needs. Soon copper was king in Butte.

By the late 1870s Butte's population was 22,000 and there were more than 300 operating mines. The population later peaked at perhaps close to 100,000, and at least for a while Butte was the largest city in the Great Plains and the northern Rocky Mountains. At the end of the nineteenth century, Butte was the largest copper producer in America and one of the largest in the world. Stock in the Anaconda Company was blue chip and included in the Dow Jones Industrial Average.

The downside of all this prosperity was an average of a death a day in the mines for thirty years, until 1925. Air pollution was horrendous. In 1884 the main smelter was established in Anaconda, thirty miles away, where a 580-foot-high stack was built to disperse the poisons, but "neighborhood" smelters continued on into the twentieth century. In 1895, a hundred citizens died during a brief air inversion that kept smelter effluents close to the ground surface. Although a colorful paradise to some, Butte, Montana, was a twentieth-century backwater to others. J. Edgar

Hoover used to banish troublemaking FBI agents to the Butte office as the ultimate punishment for failure to enthusiastically follow orders.

Colorful is a word that barely describes the history of Butte, with its battles of the Copper Kings (William Andrews Clark and Marcus Daly) and political corruption on a scale rarely seen in North America. George Hearst made a part of the Hearst fortune in Butte, which led to the newspaper chain run by his son, William Randolph Hearst. Added to this scene was the violence of the unions, often against one another, matched only by the violence of the Pinkerton police, who opposed them all. Butte has been called the Gibraltar of Unionism and may soon receive recognition as a National Labor History Landmark. Already much of the town has been designated a National Historic Landmark District.

Today Butte is a city of more than 30,000 souls residing on the steep slope of a mountain amid abandoned wooden structures located over old mine shafts. It is a town that consumed itself. After 100 years of underground copper, gold, and silver mining directly below the city itself, the Anaconda Company decided to seek lower-grade ore and moved to open-pit mining. In 1955 the Berkeley Pit was opened (figure 7.1). It closed in April 1982, but not until it had gobbled up a number of established neighborhoods bordering the downtown. Small communities such as Meaderville and McQueen vanished altogether, the land beneath them consumed by the Anaconda smelter. At one point, plans were made to extend the pit into the heart of downtown Butte, but the citizens balked and voted down the scheme.

The Berkeley Pit was kept dry by continuous extraction of groundwater by a pump that was placed 3,900 feet below the surface of the mine, well below the floor of the open pit. The groundwater table in the geologic layers beneath the earth's surface is the top of the water-saturated zone below which all rock pores are filled with water. As the pit grew deeper year by year, the water table formed a giant cone-shaped depression. Everything within the surface area encompassed by the *cone of depression* (figure 7.2) drained into the pit and contaminated the groundwater that was being pumped out.

When the pumps that were holding back the groundwater were shut off in 1982, the mine began filling at a rate of 5 million gallons per day. The cone of depression became smaller and smaller as the water level rose higher and higher.

Among the common minerals in the Butte ore body are pyrite (fool's gold) and chalcopyrite, composed of iron sulfide and iron copper sulfide, respectively. When the rock is exposed to the air for the first time, and

Figure 7.1 The abandoned Berkeley Pit mine and pit lake showing its proximity to downtown Butte, Montana. As the mine expanded it consumed portions of the town. Photo by Todd Triasted.

oxygen and water come into contact with these metal sulfides, sulphuric acid forms. The acid flows through the rock and leaches other toxic metals from the rock, including copper, cadmium, lead, arsenic, and zinc.

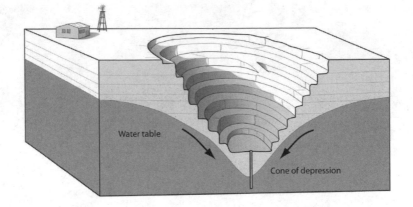

Figure 7.2 A hypothetical open-pit mine cross-section showing the groundwater cone of depression. The pumps keeping the pit dry are located well below the pit floor, as was the case with the Berkeley Pit in Butte, Montana. Arrows show the direction of flow. Once the mine is abandoned, the cone of depression gradually disappears, and the pit water, including the pollutants, mixes with the regional groundwater. Diagram by Charles Pilkey.

The same leaching process occurs when rainwater encounters the left-behind piles of mine wastes. Among the toxic metals, arsenic becomes a particular problem in waters derived from mine wastes. Arsenic comes from the mineral arsenopyrite.

The processes used to extract the copper from the ore varied at different periods of mining in Butte. In the early days when the ore was high grade, *acid leaching* with nitric acid was used, creating the acid wastes. Later, when a low-grade ore was mined from the pit, cyanide was used in huge quantities to extract copper from the native rock. The result was a strongly alkaline cyanide-laced mine waste. Whether from mine or smelter waste, it is critical to prevent acidic or alkaline waters from escaping into nearby streams. As it happens, waters from the "Richest Hill on Earth" drain into Silver Bow Creek, which then flows to the Clark Fork River, and ultimately on to the Columbia River and the Pacific Ocean.

Today the highly acidic water (pH 2.6) of the Berkeley Pit, laden with heavy metals, is more than 900 feet deep. The volume totals around 30 billion gallons. It became a Superfund cleanup site in the 1980s, encompassing 23 square miles at the surface and including 3,500 miles of underground tunnels. The Superfund status meant that the federal government had designated the site as one of the nation's major hazardous waste sites. The water in the pit is rust in color, with a hint of green that has a rotten egg odor. The Berkeley Pit and environs is one of the world's largest bodies of severely contaminated water. A 1998 *Time* magazine article refers to it as a "giant cup of poison, a man-made wonder of horrific proportions." Even so, the Butte Chamber of Commerce says the pit is the greatest tourist attraction in southwestern Montana.

The lake is now moving upward at a rate of 2 feet per month. By the year 2020 or so, the level will reach the upper and more mobile part of the groundwater, the cone of depression will be no more, and water will begin flowing into the uncontaminated groundwater layer throughout the region. This is the water used by people in town and in the adjacent valley. That lake level has been declared by the EPA to be a pit lake water depth of 1,147 feet. So all mitigation and remediation efforts are directed toward holding the pit lake depth below that number. Much less likely to occur, but certainly a lot more spectacular, is a catastrophic release of contaminated pit water down the hillside and into adjacent streams. This could happen if a fissure were produced by an earthquake or, perhaps more likely, if the water were to escape through some of the hundreds of old mine tunnels. The water could then blow its way to the surface, miles away from the pit. Keep in mind that a one-foot-square column of water in a 1,500-foot-deep mine shaft weighs close to 50 tons. It is a *Temple of Doom* movie scenario, with Harrison Ford and his companions careening along in a mine ore car, chased by a wall of water crashing through a mine tunnel. The difference, however, is that the amount and pressure of water flowing from the Berkeley Pit down a mine tunnel would be orders of magnitude greater than that portrayed in the movie.

The colossal amount of water stored in the pit is in itself a huge loss from the local groundwater system. The amount of water that will evaporate from the surface of the pit lake once it has been filled will produce a loss of water that is the equivalent of the flow of a significant stream and, of course, evaporation will continue in perpetuity.

The Berkeley Pit is also a hazard for migratory birds. A U.S. Fish and Wildlife Service report notes that birds' digestive tracts can be burned by the acidity of the pit water and natural oils from feathers can be removed, causing drowning or hypothermia. In addition, there is a "high potential for dietary toxicity through food chain bioaccumulation," a mouthful that means the birds can be poisoned. On a cold November day in 1995, the corpses of 342 snow geese were discovered in the pit. Mine officials declared that the geese died from bad grain, a theory that itself soon died.

Geese and ducks continue to be killed in the pit occasionally, despite efforts to scare them away with noisemakers.

In 1977 the mine was purchased by the Atlantic Richfield Corporation (ARCO) from the Anaconda Company, just five years before the new Superfund laws made companies responsible for cleaning up their own contamination and six years before economics forced the mine to close. The oil company was trying a new venture into hard-rock mining. For ARCO, cash rich at the time, it must have been like making a long and spectacular leap to the deck of the *Titanic* as it was pulling away from the Southampton dock. The EPA declared that the law applied retroactively. ARCO had purchased a hazardous waste site and acquired a huge financial arrearage, mostly of someone else's making.

In 2003 a new $24 million lime-hydroxide plant was constructed. The function of the plant is to precipitate toxic metals from the mine water and level out the acidity of the water at a rate of 50,000 gallons per minute. Two years later, the technology of large-scale decontamination of the waste is still in question, and another problem is that the process will produce 500 to 1,000 tons of sludge per day, which then must be stored in toxic waste ponds. If all goes well, a virtual mountain of sludge, not all of it toxic, will eventually stand beside the pit. A few hundred years from now the pollution problem will have been solved and tourists will be able to dip a drink of fresh water or catch a big lake trout right out of the pit.

ARCO vice president Sandy Stash noted that the new plant puts the company "well on the road to complete closure in Montana," but environmental groups are not so sure. Tracy Stone-Manning, of the Clark Fork Coalition, called the solution a Band-Aid approach to an enormous problem and said, "We do not know how to make the threat of that pit go away except to pump and treat that water forever."

Clearly the Berkeley Pit problem is something that the nation wants to avoid in the future. There are some open-pit mines in Arizona and New Mexico that, when finally abandoned, will have pit lakes with larger volumes than that of the Berkeley Pit. These are on private lands where the general public and the government's attitude toward a polluted pit lake is more lax than for mines operated on publicly owned lands run by the Bureau of Land Management (BLM) or the USDA Forest Service. University of Nevada mining geologist Glenn Miller argues that when viewed in the context of 100 or 200 years or more, all land should be considered public land and it doesn't make sense to leave an ever-increasing number of poisoned lakes scattered across the nation. The poisoned lakes are by-products of the immediate gain of today's mining industry, but what about our great-grandchildren? Miller notes that while underground mining is more costly, it is far less environmentally damaging, and that should be the way to go if we are guided by concern for future generations.

Another approach that holds promise for some small pits is to backfill mines with the wastes produced during mining, or with limestone, which would neutralize the acids produced in the pit waters. Miller notes, however, that this usually results in essentially the same groundwater pollution problem that a pit lake causes.

In 2004 the Newmont Mining Corporation copper-silver-gold Phoenix Mine in Lander County, Nevada, opened, with the anticipation of thirteen years of operation, followed by reclamation activities. The BLM has required the mine to be backfilled with both limestone and mine waste after operations are completed. Newmont has said that it plans to treat the waste until it is neutralized, but estimates indicate that the dumps may produce acid for 20,000 years. It is not too likely that Newmont will fulfill that obligation. Shades of Yucca Mountain! Newmont gained international notoriety in 2005, when a number of its executives were jailed in Indonesia for dumping mine wastes at sea (in Buyat Bay), causing an epidemic of mercury-related illnesses among local people.

In the last two decades a number of open-pit gold mines with sulfide minerals have opened in Idaho, Montana, Nevada, Utah, and Colorado, mostly on BLM land. Nevada alone has more than thirty open-pit mines. The Zortman/Landusky Mine in Montana produced considerable acid drainage, although none was predicted before the opening of the mine in 1979. The *New York Times*'s October 2005 series on the global environmental cost of gold mining noted that as late as 1990 a company report predicted there would be no problems from acid drainage. A mine worker who smelled cyanide in his tap water discovered the fact that cyanide was leaking into the groundwater. The mine owner, a subsidiary of Pegasus Gold Corporation, a company with most of its assets in Canada, declared bankruptcy and left the bulk of the cleanup to others. The Thompson Creek gold mine in Idaho produced significant acid drainage, which also was unpredicted.

The Brohm Mining Company, owner of the Gilt Edge Gold Mine in the South Dakota Black Hills, commenced mining there in 1989 and went out of business in 1999. The company left behind 150 million gallons of highly acidic, heavy-metal-laden pit water and millions of cubic yards of acid-generating waste rock, now declared a Superfund site. Before declaring

Figure 7.3 A 1998 photo of the operating Homestake open-pit gold mine in the Black Hills of South Dakota. The mine was a combined open-pit/underground operation, which closed in 2002. Photo from the South Dakota Department of Environmental and Natural Resources.

bankruptcy, Brohm, a Canadian company and a subsidiary of the Dakota Mining Company, extracted $69 million of gold while leaving behind a mess that will require an estimated $40 million to clean up. During its decade of operation, the Brohm Company frequently violated the environmental conditions of its permit. In 1998 alone, seven violations were recorded. In 2001 the *Seattle Post-Intelligencer* newspaper labeled the Gilt Edge Gold Mine as one of the five worst mines in the United States. The company CEO blamed the demise of the mine on the environmentalists.

As another mining company declares bankruptcy and flees north across the Canadian border, the need for some sort of prediction of acid drainage into a mine pit is highlighted in a most visible way (figure 7.3). A spokesperson for South Dakota governor Bill Janklow (by way of deflecting blame from the government) stated, "If somebody had a crystal ball, they would not have allowed gold mining there in the fist place." But maybe good science and strong enforcement would have done as well as a crystal ball.

The Elements of Pit Lake Evolution

Pit lake mathematical modeling is designed to predict the composition of the lake water long after the mine has been abandoned. It is a herculean task. The water filling abandoned pits will respond in different ways depending on the climate, the level of the groundwater table, and the oxidation state and composition of the rock that is being mined. Limestone rock is likely to produce a less toxic pit lake, since the lime counteracts and reduces acid production. Some open-pit mines don't penetrate the groundwater table and thus don't leave behind a lake. In some situations, pit lakes come and go according to seasonal rainfall. In open-pit mines that don't mine sulfide minerals (e.g., marble mines), the eventual pit lake may be close to drinking water in quality. In tropical Africa, the neighborhoods of pit lakes have become mosquito-breeding malaria centers, a particularly insidious hazard for local inhabitants.

The critical parameters that determine the composition of pit water are the following:

The balance between inflow and outflow of water from the lake. Water balance affects the composition of the lake in many ways. Water comes into the pit as rain falls on the lake surface and as groundwater seeps in from the adjacent rocks. Water leaves the lake by seeping into the rocks once the lake is full and by evaporation from the lake's surface. All of these terms are uncertain, especially in the long run of several decades and beyond. The inflow is easy to measure once the mine is abandoned and the pit is filling up with water, but how do you estimate for environmental impact statement purposes the inflow and outflow before the mine is opened?

The composition of local groundwater. With increasing time from the date of a mine's opening, accurate, meaningful prediction of groundwater quality becomes much more difficult, if not impossible. The composition of the groundwater flowing to and especially out of a pit lake usually changes significantly over time. The metal content could decrease as the groundwater is diluted by flow from outside the mining area, or it could increase as increased dissolution of metals occurs in the rock as a result of a host of chemical changes, such as exposure to air.

The pit wall contribution is often the most important source of pollutants once the lake is full. While the lake is filling, pollutants are derived from the entire air-exposed cone of depression. Of particular importance is the reaction of sulfide minerals with water and oxygen, which is responsible for production of acid. In turn, the acidic lake water interacts again with the local rock, releasing more metal pollutants.

Chemical reactions within the lake also strongly affect the ultimate quality of pit water. Reactions include those between lake water and the sediment that is accumulating on the floor of the pit as well as chemical reactions in the water itself. Reactions of lake water with the pit floor sediment are particularly complex and difficult to predict.

Acid production is the primary long-term concern in most pit lakes. The National Academy of Sciences volume on hard-rock mining points out that the variability over time in the numerous factors that control acidity limits the reliability of pit water models. The academy panel notes that acid generation is dependent upon the following parameters:

- abundance of sulfide minerals such as pyrite
- amount of iron in the lake water and the groundwater
- particle size of mine waste material
- oxygen concentration
- seasonal temperatures
- level of pollutants in lake water concentrated by evaporation
- availability of acid-neutralizing material (limestone)
- lake water pollutant saturation level
- reactions with bacteria

The Travails of the McDonald Gold Mine

In the opinion of many old-timers, things are going to hell in a handbasket in Montana. First it was those Californians moving in who have just sold their houses back home at astronomical prices and now have huge bank accounts to buy Montana land and elevate the price of property for all. Then came the movie stars and celebrities like Ted Turner, buying up big ranches. And now these newcomers are objecting to the old way of life and are even opposed to mining, the staple of Montana industry that paid high wages and made the state what it is.

What actually is happening is a new sensitivity to the environment in Montana. The evidence of the impact of mining is all around the state's residents, and the Berkeley Pit, the biggest disaster of them all, is staring them in the face. Montanans want no more Berkeley Pits. The state is looking with greater care than ever at newly proposed mines and asking tougher questions before granting permits.

One of the new mine proposals is the McDonald Gold Prospect, near Lincoln, Montana. It is a deposit estimated to contain at least 10 million ounces of gold and 54 million ounces of silver. The prices fluctuate a lot, but in early 2006 an ounce of gold was worth around $503 and going up. An ounce of silver was around $5 and going down. Within a couple of miles of the McDonald deposit there are two smaller proved gold concentrations, the Seven-Up Pete deposit and the Keep Cool deposit. Owned by Canyon Resources, the McDonald Prospect is one of the largest known untouched gold deposits in North America. The company proposed to mine by open pit and to extract gold by crushing the ore and leaching the gold from the rock using cyanide.

Cyanide heap leaching is used in most gold mine operations around the world, and a number of environmental problems have resulted. The process involves piling the gold ore in a heap and soaking it in a cyanide solution. Gold is leached from the ore and collected at the bottom of the heap. In the United States, more than $100 million has been spent to clean up cyanide and other heavy metals released to the environment by a now bankrupt mining company at Summitville, Colorado.

In November 1998 the citizens of Montana passed Initiative 137, which forbids the use of cyanide in open-pit mining operations. It was the first such prohibition by any government entity in the world and a huge blow to Canyon Resources. The vote came as a direct result of the environmental disaster of the Zortman-Landusky Mine. In February 2000 the State of Montana canceled the mineral leases for the McDonald project. Now only lawyers reap gold as the company tries to reverse the state's permit decision.

During the long process leading up to a mining permit, Canyon Resources produced a number of preliminary analyses to predict the environmental impact of the pit lake after the mine closed. The predictions were a mixture of work by two consulting groups, Shafer and Associates and Johnson Environmental Concepts. The goal was to predict what the composition of pit lake waters would be, after all was said and done, the mine closed, and the pit filled with water. This modeling, of course, was done before the bulldozer cleared away the debris from the first charge of dynamite and opened the pit.

The McDonald Gold Prospect mathematical modeling approach is typical of such exercises. Each step involves educated guesses and simplified assumptions about chemical processes. The accuracy of the numbers obtained at each stage of the process depends on how good the prediction was at the previous stage. The final number, the estimate of the toxicity of the pit water, is fragile indeed.

It's not that these numbers are pulled out of the air. The various rock types in the proposed McDonald pit were extensively sampled. A total of 609 drill holes were used to define the deposit. Groundwater flow and composition were measured in wells. Laboratory experiments were carried out to observe chemical reactions under controlled conditions assumed to emulate actual mine conditions in the future. But nothing imitates time except time itself, and changes in the rocks and the chemicals in the water over the long term will always remain in the realm of speculation.

In pit lake mathematical models such as this one, groundwater flow rates and directions are critical, but one never entirely knows the location of faults and fractures and changes in pore space in the rocks. And as chemical reactions occur within the rocks, flow character may change dramatically. Where numbers are missing, they are borrowed from other locations or other mines, always with an explanation of why this is a valid approach. It is a hugely complex physical and chemical system that controls the eventual composition of pit lakes.

The following is a generalized summary of an actual prediction carried out in happier times (the mid-1990s) for the owners of the of the McDonald Gold Prospect.

Step 1. Assume the time it will take for the pit to fill with water after mining ceases.

Step 2. Assume a water balance.

Step 3. Assume the chemical composition of the water from each of the sources comprising the water balance.

Step 4. Predict acidity by mixing (on paper) the major constituents of the waters. To do this, a model developed by the U.S. Geological Survey, *PHREEQE*, is utilized.

Step 5. Predict the metal content of the lake water by the *Latin Hypercube Stochastic* modeling approach.

Step 6. Predict the lake's physical and chemical character (limnology) using the *CE-QUAL-W2* model. Assume the lake is stratified (the Johnson consultant's report) or is uniform from top to bottom (the Shafer consultant's report).

Step 7. Predict the ultimate pit lake composition using *MINTEQA2*, a model developed by the EPA.

Step 8. Compare results to EPA water quality standards.

A chain of at least five mathematical models was used in the process of predicting eventual pit water quality in the McDonald mine. One bad link in a chain breaks it. One error in the model similarly destroys the chain.

The very first step in the process, prediction of pit filling time, is critical. In this case, it was predicted that the pit would fill in twelve years after being pumped dry during twelve to fourteen years of operation. It was assumed that twelve years after the mine halted operations, 834 gallons of water per minute would flow into the McDonald pit and be exactly balanced by outflow of the same amount.

Robert Moran, a prominent geochemical and hydrogeological consultant from Boulder, Colorado, whose views are discussed in detail below, notes that the flow models used in this part of the pit lake modeling process are mostly models or variations on models originally developed by the U.S. Geological Survey for conceptual modeling. They were designed to provide general (qualitative) guidance as to flow directions and orders-of-magnitude flow quantities. According to Moran, the models were never intended for the quantitative modeling now widely used by the mining industry.

Perhaps in an ideal world, the U.S. Geological Survey designers of the mathematical models would have been obliged to point out the misuse of the models as quantitative predictive tools. But it is usually not the nature of federal agency scientists to criticize one another unless they have been backed into a corner and forced to do so. In fact, it is in the nature of most scientists to avoid the shot, shell, bedlam, and controversy of contentious projects in the public eye. The chances are that complaints about model misuse will be restricted to grousing among fellow scientists. How different the use of quantitative mathematical models would be throughout our society if scientists routinely stepped into the public debate.

On a broader front, Stuart Rojstaczer, onetime groundwater geologist, model critic, and now a music producer, argued that we cannot accurately quantify groundwater flow at any location under most conditions because there are too many unknowns. Rojstaczer suggests, somewhat tongue-in-cheek, that we should present our results of modeling of groundwater movements by "transcribing them on the back of an envelope." He refers to the modeling of the movement of groundwater through rocks as a "digital back of the envelope" process. Putting the results on the back of an envelope would signify to the user just how risky the numbers are. Rojstaczer believes that we should inform citizens, legislators, lawyers, and bureaucrats that we cannot accurately predict the flow of water through rocks and that they need to rethink their reliance on quantitative predictions.

Tom Myers, a groundwater modeler who works for an environmental group called the Great Basin Mine Watch, is more optimistic than Rojstaczer about the validity of groundwater models. He agrees, however, that applying the models to the flow of water into giant pits dug into the middle of an aquifer goes far beyond that for which the models were originally developed. Flow of water into a pit is much different than normal flow through the cracks and pores of rocks in an aquifer. As for the models that predict chemical changes in pit lakes, Myers notes that they have never been calibrated in the field. As is the case of so many applied modeling efforts, no one has looked back.

A number of technical papers continue to show up in the mining engineering literature, almost all written by consultants who seem either blithely unaware or irresponsibly ignorant of the complexities of pit lakes. The authors seem to come from some other world, where nature is simple, predictable, and sits perfectly still while being modeled. One such 1999 paper concludes: "A simple ... methodology is presented to aid ... in predicting the ultimate water quality of pit lakes. Original equations have been developed for estimating the passive groundwater inflow rate to an open pit. . . . The method also considers the time required to reach chemical steady state given the pit lake volume, inflow and evaporation rate."

It would appear that this engineer has solved the major problems that have stymied others for decades. In a five-page paper the author expresses understanding of processes that no one else is close to solving. The very fact that such an absurdly simplistic analysis can still be proposed and published in the technical literature is a measure of a woeful state of affairs in the science and engineering of pit lake chemistry.

In 2005 mining consultants Ann Maest, James Kuipers, Connie Travers, and David Atkins presented the results of the most extensive synthesis ever of the models used to predict water quality at hard-rock mine sites (table 7.1). Although a number of uncertainties are discussed, some of them quite serious and highly likely to be insurmountable, the authors basically believe that models can accurately predict pit lake quality years into the future. It's a common modus operandi for quantitative modelers: Admit to uncertainties and complexities, yet in the end ignore them and recommend the modeling approach. It is as though admission of fatal flaws somehow erases them.

Table 7.1 shows an amazing array of models, each intended to describe one of the processes that determine water quality fifty years or more down the road. Fifteen models are listed for the calculation of water

Table 7.1 Models Used for Predicting Water Quality from the Summary by Ann Maest and Three Colleagues, of Models Used for Prediction of Mine Wastewater Composition

Process	Model	Process	Model
Water Balance	HELP SOILCOVER CASC2D CUHP CUHP/SWMM DR3M HEM-HMS PRMS PSRM SWMM TR20 SESOIL PRZM HSPF U.S. EPA 2003a	Vadose Zone (above water table)	SESOIL HELP U.S. EPA 2003b CHEMFLO-2000 Hydrus 1-D SWACROP SWIM HEAPCOV Ubsat 1 Unsat H Hydrus 2D FEFLOW Seep-W SUTRA VS2D VS2D/T
Groundwater Models	MODFLOW MT3D MODFLOW- SURFACT SUTRAFEFLOW FEMWATER FRAC3DVS FRACTRAN TRAFFRAP-WT	Lake Models	CEQUAL-W2
Stream River Models	HEC-HMS ACOE 2000 TR-20 TR-55 SWRRB PRMS SHE HEC-2 FLDWAV WASP-4 OTIS-OTEC SWMM Mike 11	Soil Erosion	RUSLE
Watershed Models	MIKE SHE SOGREAH PRMS/MMS HSPF U.S. EPA	Geochemical Reactions	WATEQ MINEQL HYDRAQL REACT PHREEQE/PHRQPITZ EQ3//6 SOLMINEQ GEOCHEM WATAIL SOLVEQ-CHILLER PATHRAC MIN3P RT3D NETPATH
Pyrite Oxidation	PYROX Davi/Ritchie FIDHELM TOUGH AMD		

balance and fifteen for the Vadose zone (water flow above the groundwater table). Fourteen models are available to address chemical reactions in the rocks. On the surface of things, the large number of models might seem appropriate. Certainly this is an improvement over the one-model-fits-all outlook in the case of beach behavior modeling by the U.S. Army Corps of Engineers. But these various models are not necessarily designed to be used for certain rock types, certain mineral assemblages, or a variety of climates. Instead the choice of model seems to be governed by such factors as custom, experience of the consultant, and available software. And since there are so many models, you can choose the one that fits the parameters you have in hand or, more likely, the one that will come up with the "reasonable" answer. Impossible models for complex processes that we clearly can never predict are piled on top of other impossible models of other complex processes.

Just as in the other modeling arenas we have discussed, accurate prediction of future water quality is a fantasy supported by a hyperreligious faith in the predictive power of numbers. After long experience in looking at quantitative pit lake models, Robert Moran characterizes the mathematical modeling community as "a priesthood; an unassailable, untouchable priesthood that speaks and preaches in Old Latin." Jim O'Malley, the fishery industry representative quoted in chapter 1, also views mathematical modeling as a priesthood: "It has developed its own language, rituals, and mystical signs to maintain its status and to keep a befuddled congregation subservient, convinced that criticism is blasphemy." Arizona State University geologist Daniel Sarewitz adds: "What is particularly disturbing about this faith is that it knows no ideological boundaries."

Mixing Mining Politics and Mining Science

In the discussion so far we have made the point that the final product, a prediction of pit lake composition, is not possible even if the taint of politics is left out of the decision process. The models fail completely to provide numbers with the certainty required by society.

Robert Moran has led the charge in opposing the misuse of quantitative mathematical models to predict the long-term composition of pit waters. He has advised corporations, concerned citizens groups, the World Bank, and even the European Union concerning the potential environmental impact of mines all over the world. Moran, perhaps more than any individual, is responsible for alerting the public, mine owners, environmentalists, and the Bureau of Land Management to the impossibility of accurate prediction of pit lake water quality. He is a strong supporter of conceptual and qualitative mathematical models but a strong opponent of the applied quantitative models used by the mining industry.

The BLM has allowed more open-pit mines to commence operation in the western United States than has any other federal agency. The agency's mission is concerned with managing vast landholdings throughout the West. Some of this responsibility was once shared with the Bureau of Mines, an agency that disappeared under the Republican Contract with America in 1996.

Moran notes that the BLM has a dilemma. On the one hand, the agency is required by law to encourage mining on federal lands. On the other hand, the agency is expected to act in the public's interest and make sure that the impact of mining on generations to come will not be a negative one. Less charitably, the BLM political context could be described as a huge and unsolvable conflict of interest.

Until about fifteen years ago the BLM environmental impact studies for new mines consisted of qualitative opinions, usually rosy and positive about future water quality of the lakes that form at abandoned mine sites. Most of the time, something close to drinking-water quality was predicted. With increasing pressure from environmental groups, however, the BLM had to make a change. So the agency turned to predictive mathematical models. Moran believes it did so in an attempt to make its predictions appear more scientific and trustworthy and to foster a "sense of certainty" about the results. Management wanted bulletproof numbers. They did not turn to models because they had any indication that the models would improve predictions. In fact, few within the agency understand the models, and consultants are hired to run the models for them.

In recent years, virtually every prediction made by the BLM models has been optimistic. Moran points out that "the usual cause of the overly optimistic scenarios lies not in the science" but rather in the government managers and the pressure they apply to the modeling consultant. In actual fact, very little is known about how these new lakes will evolve, and little research to correct this dearth of information is going on now.

The National Academy of Sciences hard-rock mining panel pointed out that pit water quality predictions are rarely revisited or checked. Much like the Corps of Engineers and its failed predictive modeling of beach behavior, the BLM does not look back. The bureau doesn't systematically attempt to learn how well the modeled predictions worked or to learn from its past errors. Of course, it may take many years before the quality of pit water is known, but that's no excuse for not trying. For both the Corps of Engineers and the BLM, it's onward and upward to the next project.

These applied quantitative mathematical models are in the wrong hands. The technical and management staff of the BLM seem not to understand the uncertainties involved. To the agencies' great discredit, they have failed to acknowledge their own lack of expertise and their own possible conflicts of interest and have avoided bringing in completely independent experts to review the mathematical model results.

The consultant problem is a difficult one in many areas where science and society mix. In a number of government agencies, the Corps of Engineers and the Bureau of Land Management being two prominent ones, it is widely recognized that a successful consultant—one who will get repeat contracts—must "work with her client to get the correct answer." Put far more bluntly, technical consultants must come up with the answer the client expects or their business with the agency could be finished.

An unwritten rule of the mining and geochemical-geological consulting community is *Thou shalt never rock the boat*. It is a cozy community, backing up one another's work very much in the fashion of the coastal engineering community in Florida. Loose cannons like Robert Moran are rare indeed.

In a chapter of the book *Prediction: Science, Decision Making, and the Future of Nature*, published in 2000, Moran illustrates the political problem of the mine environmental impact prediction process by following the process for the Pipeline Mine, southwest of Elko, Nevada. Owned jointly by two companies, the planned Pipeline open pit would be 1,000 feet deep. The pit lake would eventually be 800 feet deep, and at the time of this writing the mining is well under way. Of the thirty operating open-pit mines in Nevada in the mid-1990s, none was as deep as the Pipeline Mine and there was no local experience basis to go on.

The first 1994 environmental impact statement was founded on the work by the company's consultant. BLM then hired a "third party" consultant to review the work of the company consultant. The company had a lot of influence on the choice of the third-party consultant and in fact was responsible for paying the consultant. So the outside reviewer who should have been an entirely independent party, seeking only objectivity, had two masters, the BLM and the company itself. Both of the masters wanted the mine to forge ahead and neither really sought the independent truth.

The 1994 draft environmental impact statement for the Pipeline Mine predicted that the pit lake in the scorching Nevada desert would always be water of drinkable quality, even decades down the road. It was an absurd conclusion that presumably was agreed upon by both the company and the BLM consultant. On the face of it, the study was either massively incompetent or dishonest, a response to the economic and political pressures applied to the scientists and engineers who did the work.

The impact statement for the Pipeline Mine also concluded that the pit lake was a one-way sink of metals and thus would continually cleanse itself. Moran points out that the laboratory studies to estimate the composition of water entering the Pipeline pit from the walls involved leaching tests of wall rock using fresh rainwater, a far cry from the acidified waters that would actually flow into the pit lake.

Other wrong chemical assumptions behind the environmental impact statement noted by Moran included:

- no change of the composition of the water over time
- no concentration of chemicals in the water because of lake evaporation
- no change in water composition with depth (stratification) of lake water
- no changes in water temperature with time
- no reactions between lake water and lake sediment
- no role for microorganisms in the water or in the sediment

The large number of weaknesses, simplifications, omissions, poor assumptions, and fallacious experiments behind the models was not obvious to the non-specialist, but the drinking-water-quality prediction was very transparent. No one believed it possible.

The media and environmental groups set up a collective howl, magnified by a dispute between the Shoshone Indian tribe and the BLM. The Shoshone claimed some of the mine land as their own, and they were concerned that the water drawdown during mining would destroy their springs and wells. In response to the hue and cry, a new third-party consultant was hired, one with an impeccable reputation for straightforwardness. The consultant was Robert Moran, then with Woodward Clyde Consultants. The mining company hired a new consultant as well, and all the consultants, new and old, worked with the company's original consultant to come up with new numbers.

The new numbers predicted an alkaline pit with high concentrations of dissolved solids. Although the results were clearly more in line with reality, the modeling was still simplistic. Moran noted that the same weaknesses present in the first model were largely still present. The difference was that the modelers chose scenarios and made assumptions in the various model stages more in line with the known chemistry of sulfide mine pits. The numbers were closer to reality, but they were still quite uncertain. The irony was not lost on some of the pit lake critics that apparently it was possible to order up model results to one's particular likings. It was also apparent that the prediction could have been made just as accurately on the back of an envelope, using only intuition based on experience in other pit lakes.

It was this uncertainty that Moran insisted be recognized in the 1996 final environmental impact statement. The document included a single paragraph about model uncertainties, *over BLM objections*, which included the line "Only through future monitoring will the actual [pit lake water] concentrations be known." BLM finally admitted, albeit in some very gray literature, that it was not dealing with the "gospel truth" in an environmental assessment.

Robert Moran's revelations concerning the Bureau of Land Management's mine waste modeling problem illustrates most of the same issues we have seen in other arenas of modeling. In particular, pit lake modeling is a prime example of the political vulnerability of model predictions to distortion, as the truth is molded according to a client's needs. We see in this type of modeling also what can happen when models are used by those who have no understanding of them, no knowledge of their uncertainties, and no intuition about the science behind them.

The Foreign Connection

There seems to be little question that the current U.S. practice of expecting accurate characterization of the ultimate composition of a pit lake after a mine has been abandoned as impossible. As open-pit mining accelerates around the world (because underground mining is often more costly), the problem of pit lake prediction is now a global one. As it becomes increasingly difficult and costly for new mines to open within the United States, more companies are looking to the international scene. In the 1960s and 1970s a lot of overseas mines were caught up in a wave of nationalizations, but now the global business scene is a more favorable one. American mining companies are welcomed again.

For example, Meridian Gold, co-operator of the Beartrack open-pit mine, which proved to be a major polluter near Salmon, Idaho, is now seeking to open the El Desquite open-pit gold mine in Esquel, northern Patagonia, in Argentina, a country desperately in need of economic development. Robert Moran advised local government in Patagonia about the pitfalls of pit lake quality prediction, and the local people rejected the mine. The company continues exploratory drilling, however, apparently confident that the situation will turn around in its favor.

Moran notes a number of common problems concerning the evaluation of open-pit mine environmental impacts in developing countries.

- The host government is usually a business partner with the mining company. Thus there is little particular incentive for finding the truth about environmental impacts.
- The ministry that promotes mining also regulates mining (largely true in the United States).
- Historically there are no consequences for a mistake made by the mining company (largely true in the United States).
- As a rule, very little actual data are available with which to make a prediction about environmental impact.
- Favorable economic and environmental predictions are good for stock prices. One company's stock rose 120 percent when favorable project impact predictions were announced for a proposed mine in Tambogrande, Peru.

The Solution?

Halting open-pit mining would solve the problem immediately. Alternatively, at the other end of the spectrum, we could simply let mines open and learn to live with polluted pit lakes in our midst. Of course, in doing so, we would be saddling future generations with a landscape that is littered with many cups of poison. Neither of these scenarios will be acceptable to our society, so some sort of accommodation with the environmental impact of open-pit mining is a necessity.

Glenn Miller's "solution" is to encourage or require underground mining, which he describes as orders of magnitude less damaging to the environment than open pits. But what to do about the increased costs of underground mining? Can such mines still compete on the international scene?

Moran's "solution" to the impossibility of accurate predictive modeling and the lack of independence in the BLM environmental impact assessments is to require the mining companies to deposit large bonds to repair future pollution problems. The idea is to post financial bonds

Figure 7.4 The beginning of infill reclamation of the Landusky Gold Mine in the Little Rocky Mountains in north central Montana. This mine, operated by Pegasus Gold, has produced significant acid drainage into the local groundwater. Photo from the Montana Bureau of Mines.

to ensure perpetual maintenance and pollution protection long after a mine closes. From Moran's standpoint this is one way to stop the mathematical model charade.

The State of Nevada's Commission of Mineral Resources requires companies to participate in a reclamation bond pool (figure 7.4). In early 2003, however, the total amount in the pool was less than $2 million, hardly enough to clean up a single major sulfide mine pit, much less dozens of them.

Recent events in Nevada and other western states provide another argument in favor of bond posting. No less than thirty-two Nevada mines have gone bankrupt since 1998. If companies disappear altogether, no one remains behind to clean up the mess except the deep pockets of Uncle Sam. In the Payette National Forest in Idaho, the Dakota Mining Company abandoned its stibnite (antimony) mine, forfeiting an $800,000 bond to cover a multimillion-dollar cleanup. The aforementioned South Dakota Gilt Edge Gold Mine run by the Brohm Company, a

Canadian subsidiary of the Dakota Mining Company, left behind a $5.6 million bond to cover a $40 million cleanup.

But the bond idea doesn't meet with universal approval. The proposed Crown Jewel Mine in northwestern Washington State was slated to be that state's first open-pit mine, until the project was dropped in 2001. The State of Washington's Pollution Control Hearings Board denied the permit, saying, "The [bonding] approach is tantamount to entering a busy interstate highway on an exit ramp against the traffic. The availability of insurance in that circumstance is no more comforting than the proposed bonding here."

The evolution of abandoned open-pit mine lakes is clearly just as complex as beach behavior, sea-level rise, fishery evolution, and any other physical, chemical, or biological process at the earth's surface. Sheltered by the fig leaf of sophisticated and state-of-the-art mathematics, however, the BLM and other agencies have not looked back in any systematic way.

Tempting as it will be to government bureaucrats to continue the use of models, the predictive models for the long-term quality of water in abandoned open-pit mines should themselves be abandoned. And after future environmental impact statements are made, the permits granted, and the mines opened, close monitoring and strong enforcement of mining regulations would help greatly. Because of the cups of poison that industry leaves behind itself, the time has come for society to contemplate the future of open-pit mines.

The future is an opaque mirror. Anyone who tries to look into it sees nothing but the dim out-lines of an old and worried face.

—*Jim Bishop*, New York Journal American, *March 14, 1959*

chapter eight

invasive plants

an environmental apocalypse

Birds don't sing in the jungles of Guam. A walk through this tropical paradise is a walk in silence. The quiet is caused by a snake, the brown tree snake (*Boiga irregularis*). If ever there was an obnoxious invasive or-ganism, this is it (figure 8.1).

The list of this snake's loathsome characteristics seems endless. For starters, this thin, seven-to-ten-foot-long creature is said to be more ag-gressive, though less venomous, than a rattlesnake. It injects its venom while chewing on the victim. It seems particularly fond of newborn mammals, such as puppies and bunnies, and has taken a liking to poul-try. The impact on poultry and egg production in Guam is significant, necessitating importation of eggs from other islands at increased cost to the local population. There is an absence of natural enemies, and in the half a century since its introduction, the snake has almost destroyed Guam's native population of small mammals and birds, which in turn has had a huge impact on plant pollination and insect control. Twelve bird species have disappeared from Guam. Two of these species, the Micronesian kingfisher and the Guam rail, once existed only on Guam. They now exist only in zoos.

Figure 8.1 The brown tree snake of Guam. These poisonous snakes, seven to ten feet long, have essentially wiped out the bird population of Guam. Plants have been affected because of the loss of the role of birds in spreading plant seeds. Photo courtesy of the U.S. Geological Survey.

At least 200 people, many of them babies, have been bitten in recent years on Guam. The snake's most startling behavior has been to attack babies as they lie sleeping in their cribs. It crawls up and through and around almost anything. In one instance a snake injected poison into the hand of a baby lying in a crib and then tried to swallow the hand. The baby survived.

The snake population is around 13,000 individuals per square mile. Because they seem attracted to power lines and transformers, these snakes cause frequent power outages as they spectacularly electrocute themselves. As the snake slowly silenced the jungles of Guam, it changed its food habits to feed primarily on other introduced species, particularly the curious skink, a small lizard. The curious skink has become so important in the snake's diet that the snake has adapted to the skink's behavior and begun to appear in daylight and on the ground instead of in its normal nocturnal tree habitat. The snake changes its color according to food and habitat.

An ability to adapt to changing conditions is a key characteristic of successful invasive plants, animals, and insects. The change in diet and eating habits has increased the mortality of the adult snakes because they are more visible to humans as they cross yards and busy roads in broad daylight, but the introduced skinks are sufficiently abundant to keep the snake numbers high.

The snake is native to northern Australia, New Guinea, and Indonesia. It is not a particular pest in those places because natural predators there, among them monitor lizards, birds, and several mammals, keep its numbers down. Although there is uncertainty as to its arrival date, the snake probably came to Guam from New Guinea before 1950, possibly in military supply shipments during World War II. By 1970 it had become an important component of the Guam ecosystem.

The snake has changed the plant ecology as well. Loss of the birds has meant a loss of the seed dispersers for many native plants. The change in birds has also affected the insect population, which in turn has changed the pollination process for certain plant populations. Insect-eating birds are now so depleted that newly introduced insect species are thriving. Spiders seem to be playing the former roles of birds.

The news is not all bad, though. The loss of the birds has reduced the proliferation of lantana, a seriously invasive plant pest whose seeds are spread almost exclusively in bird droppings.

What has happened here is expected to eventually happen in coming decades on other Pacific Islands as well, especially the neighboring Marianas and Carolines. The snake has been spotted in small numbers on Saipan, Tinian, Rota, Kwajalein, Wake, Pohnpei, Okinawa, and Diego Garcia. Sooner or later the snakes will come to Hawaii. Some believe they are already there—there have been eight sightings on Oahu. The main defense against snake invasion is extraordinary alertness at airports in Guam as well as at all destination airports. Dogs trained to sniff out the snakes are the backbone of a huge anti-snake effort in Hawaii. One snake was found alive on a Hawaiian airport runway, apparently having ridden in the plane's wheel well and survived the lack of food and water, low oxygen, and cold temperatures.

John Berry, of the U.S. Fish and Wildlife Service, calls this "the most significant environmental threat to the Hawaiian Archipelago, bar none, in this century." Looking at a bigger picture, Tom Stohlgren, a biologist with the U.S. Geological Survey, in a 2005 interview with the *Washington Post* held that invasive plants and animals are the greatest environmental threat of the new century, even greater than global warming.

Ironically, most of the spectacular plant or animal invasions around the world have occurred because of blunders by people, missteps that could have been prevented by a little homework. Take the 1935 introduction of the cane toad to Australia, for example. Intended to eradicate crop-damaging cane beetles, these poisonous toads, which kill small marsupials and even toad-eating crocodiles, hopped about on the ground while the beetles remained safely out of reach high up on sugarcane stalks. Meanwhile over the ensuing decades the toads have hopped and hopped and spread for hundreds of miles.

An even more famous introductory failure was the arrival of the small Indian mongoose to the Caribbean Islands. Famous for killing cobras in India, these animals seemed to be the perfect way to rid Martinique and St Lucia of the deadly Fer de Lance snake. Legend has it that this vicious snake was introduced, at least on Martinique, to keep slaves on the plantations. But the snakes proved to have no racial prejudices whatsoever and menaced slaves and slave owners alike. The problem with snake eradication was that the mongoose fed during the day and the snake fed at night. So the two rarely met.

The mongoose was also introduced on most other Caribbean islands to rid cane fields of rats. But the rats proved to be too smart and the mongoose failed to dent their populations. So there was nothing left for the mongoose but to eat bird eggs, small mammals, chickens and eggs, small dogs and cats, and even a few snakes and snake eggs. On some islands, the population of indigenous animals has been devastated by the mongoose, which never did do what its introducers thought it would.

Even the earthworm, the common fishing variety, which is widely thought to be a beneficial organism, is a bad actor. Most earthworms are invasive species from Europe and Asia; the native worms having been wiped out in North America by the cold weather that accompanied the ice ages. Although the Department of Agriculture hasn't recognized it yet, someday efforts may have to be made to halt their spread. The problems arise on relatively pristine forest floors where earthworms change the soil chemistry and biology (e.g., the insect communities) and cause native plants to be replaced by other species. In Minnesota for example, because of the impact of the worms on soils, oak forests can be overrun by invasive buckthorns.

In a 2003 paper Richard Mack, a leading invasive plant expert form Washington State University, listed some of his favorite strange and unpredictable ways that humans disperse plants. During the World War II campaign in northern Finland, both sides imported hay for horses from all over Europe. As a result, immediately after the war 140 non-indigenous plant species were identified in the battle zone. Most of the species disappeared after a few decades because of unsuitable climate. Japanese delegates to an 1884 cotton exposition in New Orleans introduced water hyacinths to the United States, the plant that clogs many warm-water rivers. The water hyacinth, however, is not a native of Japan. The Japanese delegates picked it up in Venezuela on their way to the United States and proudly gave their American friends specimens of the pretty plant. The Bathurst burr arrived in Australia tangled in the tails of horses imported from Chile in the 1840s. Sixty years later, the burr provided the body of popular paper-winged-butterfly lapel ornaments that stuck to clothing. Eventual disposal of the burrs undoubtedly helped the spread of the invasive plant.

Another example cited by Mack as the most bizarre form of seed dispersal he is aware of involved a Mr. Hillman, a Nevada agricultural expert who was concerned with invasive plants in the 1890s. Hillman devised pamphlets describing plants to watch out for and included in each pamphlet an actual dried specimen of the plant *with seeds*. He widely distributed the pamphlets to farmers and thereby undoubtedly distributed the invasive plants at the same time!

An Out-of-Place World

Biological invasions of plants, animals, insects, and pathogens are a global threat to ecosystems everywhere. This is a widely recognized threat, one that we read about almost daily in the news. Who hasn't heard about the rabbits in Australia and kudzu vines in the American South? What gardener hasn't sputtered in frustration about johnson grass and Chinese bamboo grass taking over the vegetable plot? Or the Asian silver carp that has brought water skiing to a screeching halt on the lower Missouri River? The fish, which can weigh over fifty pounds, jumps out of the water in front of or behind speeding boats and slams into skiers and boaters.

And that's not the half of it. The giant hogweed, a beautiful towering bush, came from Asia to Britain in Victorian times and from Britain to the United States in the last century. It is carcinogenic and causes skin problems that make poison ivy seem rather insignificant. The kudzu-like Old World climbing fern from Australia and Africa spreads by air, water, and soil and may yet take over the Everglades and much of South Florida. The yellow star thistle, an invader from Turkey and Armenia, is now the most common plant in California. It is poisonous to livestock and is a prickly plant that has ruined pastures and hiking trails and displaced native plants and animals. It has caused huge economic losses to ranchers.

A *non-native species* is one that is found in an area in which it did not evolve; it usually got to that ecosystem with human help, whether intentional or not. Such organisms are also called *exotic, alien,* or *non-indigenous* species. Most non-native species cause no apparent harm to the environment. It is only when a non-native species is very successful and threatens the survival and well-being of native species that it is considered to be *invasive. Naturalized plant species* are non-native species that have successfully established themselves, usually at a low level of abundance, without being invasive. *Weeds* can be either native or non-native species that spread and hamper the growth of crops or other desired plants.

Invasive organisms include animals, plants, and pathogens. In this chapter plants will be emphasized. Pathogens, along with arthropods, can be collectively referred to as *plant pests. Pathogens* are any microorganisms or viruses that can cause plant diseases. *Arthropods* are invertebrates that have segmented bodies and jointed legs and include insects, spiders, and centipedes. The introduction of pathogens and arthropods to an environment is generally accidental, facilitated by their "hide and survive" capability. Container shipping and very rapid air transport of people and cargo are boons to the introduction of non-native plant pests and pathogens. Plants and animals, on the other hand, usually don't arrive in the United States by accident. People bring them in.

Plant species can be transported by natural pathways, too, but these routes are much less efficient than human pathways. Natural transfer can take place over very long distances, as evidenced by isolated Pacific islands with luxurious flora, containing some of the same species found on continents thousands of miles away. Alert beachcombers anywhere in the world can spot plant remains that come from far away and from vastly different climates. For example, on U.S. East Coast beaches, coconuts, mangrove seeds, tropical logs, and other out-of-place plant remains are reasonably commonplace. Arthropod and pathogen redistribution is helped along by migrating birds, insects, and animals and by passive transport in air and water currents.

Plants, insects, fungi, and viruses are clearly transported by wind, often during storms. For example, the pink hibiscus mealybug from Asia recently appeared in the Caribbean after a hurricane, and some future hurricane is expected to carry it to South Florida. The bean golden mosaic virus appeared in Homestead, Florida, after Hurricane Andrew (1995). In this case, winds carried whiteflies that carried the virus. Soybean rust from Asia arrived with Hurricane Ivan (2004).

Dust has probably transported organisms to new sites since the time the continents first formed, billions of years ago. African dust blowing across the Atlantic is a natural pathway for introduction of new species and an excellent example of some of the complexities facing those who would predict the future of invasive pathogens. The dust is heavily laden with bacteria, fungi, and viruses, some of them pathogens for plants. For thousands of years, but only recently recognized, dust from giant African desert dust storms has ridden the trade winds all the way across the near-equatorial Atlantic to the Bahamas, South Florida, the Caribbean, and even the Amazon rain forest. Early inhabitants of the Bahamas made pottery from the clay of African dust that accumulated in cracks and crevices of Bahamian limestones. African dust has been with us a long time.

Environmental scientists suspect that the increasing amount of dust arriving in the New World is the result of accelerating African aridity and desertification, beginning in the mid-1960s. It has probably been a factor in the degradation of coral reefs in the Western North Atlantic. One possibility is a transported fungus, *Aspergillus sydowii*, that is known to attack corals. African dust particles have recently been suspected as the cause of increased asthma in Barbados residents. Gene Shinn, a geologist with the U.S. Geological Survey, believes that there is a possibility that mad cow disease could be spread in this fashion.

It is clear that humans cause far more species displacement than does nature. The history of plant introduction to isolated islands like Hawaii provides a perspective on the relative rates of natural and human-induced plant introduction. This was the subject of the classic 1967 book by Robert MacArthur and Edward Wilson titled *The Theory of Island Biogeography*. For comparative purposes, the rate of introduction of plants to islands is expressed as the number of years between species introduction over the entire available time span, assuming their introduction is evenly spaced in time once the volcanic islands formed, beginning 5 million years ago. The plant assemblage that is now considered to be native flora could have been derived from an introduction rate of only one species every 100,000 years over the millions of years that the islands existed. When Polynesians settled on the islands 1,500 years ago, they brought several dozen species of plants, which upped the rate of introduction to one new species every 50 years. Come the Europeans, about 270 years ago, and the rate skyrockets to 22 per year, involving a total of 5,000 new plants. Insects once appeared in Hawaii at a rate of one species every 175,000 years, but now 15 to 20 new species become established in the islands every year.

Indications are that all over the world, non-indigenous species are being introduced at an increasing rate. This comes at the same time that native plant communities are under stress from fragmentation, pollution, climate change, and changes in water and fire regimes. Methods for getting rid of alien species are being halted because of justifiable concern about potential collateral damage from pesticides and herbicides. The costs and amounts of damage are increasing, and international control of invasive species has become, according to some economists, a weakest-link phenomenon. If one country or even one farm fails to take the necessary steps to halt an alien advance, then it doesn't matter how good the efforts of others are. One bad apple spoils the barrel, as the old saying goes.

Some believe that the biggest source of North American invasive plants in the future will be China. Americans look upon Chinese plants as mysterious, beautiful, and, above all, desirable. Since there has been a huge surge in trading activity between China and the United States, new species are bound to be crossing our borders. In addition, China and the United States share similar climates and growing conditions over large areas of land. Chinese tallow, Chinese privet, Chinese lespedeza, Chinese empress tree, Chinese silvergrass, Chinese banyan, Chinese brake fern, Chinese wisteria, Chinese aralia, Chinese packing grass, and Chinese yam—all are examples of invasive plants introduced mostly as ornamentals, a number of which also have folk medicinal attributes.

Chinese tallow probably arrived in North America in the 1700s. The plant has a number of traits that make it a natural for invasiveness. Its seeds can be spread any number of ways, such as by birds, especially pileated woodpeckers and grackles (which are themselves non-natives), running water, packing material, or mud on the soles of shoes of globe-trotting horticulturists. In addition, the seed of the plant remains viable for many years.

In 1772 Benjamin Franklin sent some seeds to a friend in the Georgia Colony, noting in a letter, "I send also a few seeds of the Chinese Tallow Tree which will, I believe, grow and thrive with you. Tis a most useful plant." Today the plant is found from the coastal plain of the Carolinas to eastern Texas and on the entire Florida peninsula, where it is threatening the existence of some of Florida's few remaining virgin forests. Chinese tallow is still for sale at plant nurseries.

In a fourteen-year experiment in a Texas field it was observed that tallow trees from a stock that had been in North America for many generations were more vigorous than recent Chinese imports. This has been

Figure 8.2 Spartina alterniflora, the salt-marsh grass from the East Coast, being mowed and crushed. This invasive plant arrived as wrapping material for oysters and has succeeded in wiping out large sections of the broad mud flats of Willapa Bay, Washington, to the detriment of the commercial clam industry. Note the condition of the surface of the marsh as the invasive marsh grass is being removed. A lot of mud-flat organisms are killed by this process, as well as by the application of herbicide. Is it worth it? Photo courtesy of Kyle Murphy and the Washington State Department of Ecology.

observed in a number of invasive plant species that appear benign when first introduced. Some become naturalized but don't spread for decades. As their populations build and as generations pass through, they are likely to adapt only too successfully to their new home, just as the snakes of Guam came out of the night and out of the trees and began to hunt on the ground in daylight for their new source of food. This is another of the unpredictable complexities that make predictive quantitative mathematical modeling of the future of an introduced plant essentially impossible.

The transfer of the salt-marsh grass *Spartina alterniflora* a few decades ago from the East Coast to the West Coast best illustrates the *packing material* mode of invasive species introduction. It all occurred because East Coast oysters were shipped to restaurants in Washington State, packed in Spartina straw. Upon arrival, the oysters were removed and the packing thrown over the railing into the nearby mud flat. The new grass took hold, and now a battle royal is being fought on the vast mud flats of Willapa Bay, Washington. Spartina is being yanked out and bulldozed (figure 8.2), covered over with mud, and poisoned by herbicides (figure 8.3). All of these efforts are probably to no avail. The problem is that Spartina is wiping out the indigenous mud-flat fauna, including commercial clams, and replacing it with an entirely different fauna. Interestingly enough, local citizens are divided into two vociferous groups on the Spartina issue. One group argues that the salt-marsh grass is beautiful and should be left to spread throughout the bay. The other group favors extinction of the invasive plants.

The number of bad deeds done by invasive organisms provides an almost endless list. Total extinction of native plants and animals, such as the brown tree snake brought about on Guam, is not common. But reduction in the abundance of native species is. For example, fire ants have greatly reduced the number of native ants in Texas. From Australia in 1895 came the Australian pine, a beautiful tree that now has a firm foothold in South Florida. The problem is that it nudges out native plants,

Figure 8.3 Salt-marsh grass being sprayed with herbicide in Willapa Bay, Washington. Application of herbicide is the principal basis of opposition to Spartina eradication by some local citizen groups. Photo courtesy of Kyle Murphy and the Washington State Department of Ecology.

Figure 8.4 Kudzu, sometimes referred to as the "vine that ate the South," is an invasive plant from Japan. In this photo the kudzu has swallowed an abandoned service station in Hillsborough, North Carolina.

allowing no other plants to grow in its immediate vicinity. When the hurricane winds blow, it is the first tree to fall across important road intersections in use by citizens fleeing the storm conditions.

Other famous bad actors are the rabbit in Australia, the zebra mussel in the Great Lakes, and the most famous of all, the kudzu plant in the southern United States. Known locally as "the vine that ate the South," it was introduced from Japan in 1876 at the centennial exposition. Gardeners, intrigued by the large leaves and sweet-smelling flowers, snapped it up. Today the plant covers 2 million acres of southern countryside. Until 1953, it was feted and promoted by governments and botanists alike. Then, in 1972, it was declared a weed.

Some towns in the kudzu belt have learned to roll with the punches. A sign in Chipley, Florida, proudly notes "Kudzu Developed Here." Holly Springs, Mississippi, has the annual Kudzu Festival, where Miss Kudzu is selected and citizens vie to show off their kudzu jelly, kudzu baskets, and kudzu furniture.

During the 1930s, thousands of young men were employed by the Civilian Conservation Corps to plant kudzu as an erosion-control cover throughout the South. In the 1940s farmers were paid by the acre to plant kudzu in their fields. A common story heard throughout the South (perhaps a kind of rural "urban" legend) tells of the family who buys a piece of property and later discovers there is an old house, garage, or swimming pool on it, completely covered by the vines (figure 8.4).

How to kill kudzu? Cutting it only encourages its growth. New plants will rise from piles of cut vines. Herbicides must be applied for up to ten consecutive years and are only partly successful.

Since September 11, 2001, the threat of invasive plants, pathogens, and insects has become a front-line national security issue. The brief anthrax scare caused by mysteriously mailed envelopes of anthrax powder clearly revealed the ease with which pathogens could be spread. The mad cow disease outbreak in Great Britain illustrated how the occurrence of an easily introduced and spread disease in a very small number of individual animals could be devastating to an entire agricultural economy. In 1999 an encephalitis virus struck the pig population of the Malaysian state of Negri Sembilan, forcing the government to cull 800,000 animals. This incident caused pork prices to bottom out, significantly undermined confidence in the government, and for a while was a real threat to societal stability.

Although terrorism is not suspected in either the mad cow or the encephalitis incident, the terrorism potential is clear. An example of a plant pest with terrorist potential would be any major plant pest or pathogen of an agricultural crop, such as the Mediteranean fruit fly in California.

The threat of bioterrorism is here to stay, after all, and the study of invasive plants and plant pests was suddenly upped a big notch in terms of importance and funding. Prediction of where and when bioterrorism might occur and what impact it might have has become a high national priority overnight.

Where, When, and How Fast

The status of predicting plant invasions is summed up in the 2002 National Academy of Sciences publication *Predicting Invasions of Nonindigenous Plants and Plant Pests*. The committee that wrote the book, headed up by Richard Mack, consisted entirely of academics. The report is a gale of fresh air relative to much of the literature on applied biological and geological mathematical models. It is a striking document because of the caution and concern repeatedly expressed about the possible pitfalls of quantitative mathematical modeling of the future of non-indigenous plants and the strict standards to which modeling is held. The panel suggests that "a scientifically based predictive system [a quantitative mathematical model] for invasiveness should meet three criteria" (8–9):

- It must be transparent and open to review and evaluation.
- Parameters used in the models must be based on field observation and/or experiments or both.
- The process must be reproducible and all who use it must come up with the same answer.

The report concludes that we don't have a comprehensive and dependable way to predict where, when, or how rapidly a non-indigenous species will become established in a new range. The report notes five general approaches to predicting the behavior of invasive plants and pests that are being globally applied with varying degrees of success:

- expert opinion
- climate matching
- ideal traits
- mathematical models
- pest risk assessment

Expert Opinion

Most prediction about invasive plants today is based on *identification of species that already have a record of invasiveness*. This remains the most accurate predictor. The basic idea is that if a plant, pathogen, or insect has been a bad actor when it invaded in one location, it can be expected to be the same in other locations. For plants, one approach is to look up the species in question in published lists such as *The World's Worst Weeds*, by Leroy Holm and associates. The book provides an empirical prediction as to which weeds will be invasive.

Any expert opinion approach naturally has some problems, not the least of which are the subjective nature of expert judgment and the fact that experts, given the same information, may disagree. And, of course, nature provides many complications that may baffle even the experts, such as the following:

- Some species considered likely to be invasive have not proved to be so when introduced to a new location. There seems to be a high degree of chance in the plant migration business.
- Some pathogens and insects that are harmless in their native habitat and are thus ignored have proved to be bad news upon entering a new region.
- Some plants species don't become invasive for many years after their introduction.

Climate Matching

This form of modeling, often based on the CLIMEX model, is a first-cut qualitative evaluation of the possible climate range of an invading plant. CLIMEX consists of a database of 2,500 locations around the world, including temperatures, humidity, and rainfall. The model answers two questions. One is a simple comparison of two locations: Does location A have the same climate as location B? The other question is, Knowing the range of a particular species, where else might the species exist? Major limitations of CLIMEX include:

- It is based on the wrong assumption that climate is the only predictor of a species' new range. The climate approach does not take into account such biological factors as host plants, competition, predators, and so on.
- It does not take into account chance dispersal. The native habitat of johnsongrass (*Sorghum halepense*) was subtropical North America. Today, after widespread dispersal by the activities of humans, it has extended its range into southern Canada.
- Climate in the model is characterized rather woodenly. It is made up of averages, which do not accurately account for daily to seasonal variations and extreme conditions, a problem discussed in chapter 4.

Ideal Traits

By study of the history of invasive species it is apparent that there are certain ideal traits for invasive plants and plant pests that lead them to become invasive. For example, the pathogens that cause potato blight and wheat stem rust have high spore production rates and short development times. Soybean rust is caused by pathogens with very efficient long-distance dispersal of spores. The potato wart is caused by a pathogen that can exist in soil for a long time with no host plants.

One would expect the following traits in pathogens to favor invasiveness:

- quick maturation
- a long infectious period
- a high rate of production
- efficient long-term dispersal
- a large survival rate between seasons
- a large number of host species

The following plant traits favor invasiveness:

- rapid development to reproductive maturity
- rapid and abundant reproduction
- small and easily dispersed seeds
- tolerance to a wide range of weather and other environmental conditions

The weakness of the ideal traits approach is that "good" traits are only part of the picture. For example, the interaction of the plant and plant pests with a specific environment is not considered.

Mathematical Models

Steven Higgins and his coworkers from South Africa suggest that it is possible to predict rates and directions of alien plant spread by quantifying the processes that disperse them. This type of simulation requires a number of analytical equations to describe and predict each important process. They developed the *SEIBS* model (*Spatially Explicit Individual Based Simulation*) and applied it to two species whose spreading history and distribution were well known in the South African "Mediterranean shrublands." The plants were a pine tree and a shrub, both of which had high rates of seed production and both of which had wide ranges of environmental tolerance. In fact, the range was wider than the range of environmental variability within the South African study site.

In the introduction to their paper, Higgins and his colleagues make a most important statement regarding the complexities of modeling natural processes. It reflects a level of understanding and recognition concerning mathematical models of natural earth systems that we wish was present in other fields where modeling prevails: "Generating confidence in process models is not straightforward: ecological systems are complex and modeling them involves parameter estimation, assumptions, abstractions and aggregations. This means that a modeler could make a great number and many types of errors in constructing a process model."

The SEIBS mathematical model assumes there are five key processes:

- rate of seed production
- plant mortality
- seed dispersal
- rate of plant growth
- disturbance (fire, cultivation, and so on)

Each of the processes is modeled with simple analytical equations supported by field data, which are combined into an overall algorithm describing plant dispersal. The fit between modeled and actual spatial distribution of the two plant species was considered statistically valid, but the authors take pains to point out that this was a short-distance, short-time study only and that the model approach would vary from species to species. In other words, this is not a general model; it is a highly case-specific model, and in this instance it was not used to predict spread of plants. It was used to match known plant behavior, a most important distinction.

Neither the SEIBS model nor any other such model seems to have found significant application in the real world of invasive plant studies and policy. The reason is clear. Accurate prediction of the spread of invasive plants is just too complex.

Pest Risk Assessment (PRA)

Both Australians and Americans rely on this approach as the first line of defense against invading weeds. The Australian Quarantine and Inspection Service (AQIS) uses a list of forty-nine questions, to be answered yes/no/don't know, to provide an overall score measuring the risk potential of new plants. It is called the *Pheloung System*, after one of its developers. The American Animal and Plant Health Inspection Service (APHIS) uses similar "likelihood models" to regulate plant imports. Both the

Australian and the American predictive systems are transparent to critics and easily understood by non-experts.

APHIS makes qualitative assessments of each of the parameters involved, using terms such as *high, medium, low,* or *negligible* rather than actual numerical values. For example, the categories ranking the climate and habitat suitability for establishment of non-indigenous plants are defined as follows:

- High—Score 3: Most of the United States is suitable.
- Medium—Score 2: One-third to two-thirds of the United States is suitable.
- Low—Score 1: One-third or less of the United States is suitable.
- Negligible—Score 0: There is no potential to survive in the United States.

The scores from this and similar analyses of other parameters are combined to produce an overall measure of risk and a basis for controlling invasive plant importation.

An Imperfect Science

No fudge factor coefficients or hidden assumptions here. What a contrast to many of the other modeling communities that we have reviewed in this volume. In our survey of a number of specialties that use mathematical modeling to predict how earth surface processes will unfold, the students of invasive plant behavior are a positive endpoint. These biologists seem to have achieved a strong sense of the complexity of their system and somehow have not been lured down the path of spiffy quantitative modeling. Although many plant ecologists are pursuing quantitative modeling, it remains a fringe activity that is not central to answering the critical societal questions about invasive weeds, pathogens, and insects.

We can speculate about why this group of scientists has not gone the quantitative route of pinpoint answers. Certainly it must have been a temptation. After all, other scientists claim to have been solving societal problems for decades using highly sophisticated and state-of-the-art quantitative mathematical models. Perhaps the complexity of predicting plant futures with accuracy is so huge and so obvious as to discourage even the most die-hard mathematician. Who could have predicted that a snake would halt the spreading of the invasive plant lantana in New Guinea by killing the birds that spread the seeds? What modeler would have factored in African dust as a source of pathogens in South Florida? Who would have guessed that johnsongrass would expand from its native subtropics to the subarctic. The many paths of recruitment, dispersal, growth, relation to climate, mortality, and the interactions with a large biological community in both the native and the receiving environments simply overwhelm any reasonable quantification of the system to seek answers to the questions of where, when, and how much. And if all that isn't bad enough, human behavior has to be thrown into the mix as well.

Similar complexity in other earth surface environments, such as beaches, however, did not similarly retard quantitative mathematical modeling. The beach-modeling scene is almost entirely in the hands of engineers who, lured by their success in using models of steel and concrete, believe that nature can be modeled just as well.

Unfortunately, the qualitative modeling approach by risk assessment for invasive plants has not solved the problem. We can be assured that highly undesirable plants and related pests will arrive in the future, some perhaps in the hands of terrorists. Nonetheless, the qualitative approach is the most flexible and most accurate way to go for the study of invasive plants. The transparency of this approach even allows us to spot our mistakes, to understand where we went wrong and to tighten up our future analyses. Meanwhile, quantitative or qualitative, it is certain that prediction of future invasions of plants, arthropods, and pathogens is an imperfect science. In fact, ordering complexity determines that accurate prediction of plant invasiveness will never be possible.

How could there be so much pretense, so much delusion, so much auto suggestion; why play such laughable games?

—*Vaclav Smil, University of Manitoba professor and author of* Energy at the Crossroads, *in reference to quantitative mathematical models used to forecast energy trends*

chapter nine

a promise unfulfilled

The Haff Principles

In the preceding chapters, a number of modes of mathematical model failure have been recognized, the sum total of which points to the virtual impossibility of accurate quantitative modeling to predict the outcome of natural processes on the earth's surface. Physicist-turned-geologist Peter Haff has categorized the fatal flaws in such models, most of which we have at least briefly touched on earlier in this volume. Haff specializes in the study of desert pavement, the rocky surfaces that form upon long exposure of the desert surface to sun, wind, and alternating freezing and searing temperatures. He has long perceived the futility of quantitative mathematical modeling of natural processes even as he studies the evolution of desert landscapes with qualitative mathematical models. The four most salient problems are the following:

- errors in characterization of the processes being modeled
- omission of important processes

- lack of knowledge of initial conditions
- intrusion of forces that influence events from outside the system

One of the most common sources of quantitative modeling error is *inaccurate characterization of the processes being modeled*. For example, averaging parameters in modeling is always necessary to reduce the size of the databases. Hence numbers used to describe environmental conditions, whether they be wave heights, water and atmospheric temperatures, sand grain size, groundwater flow rates, or fish abundance (among many others), are of necessity expressed as *averages*. But averages are a wooden, clumsy way to characterize nature. Nature herself doesn't deal in averages.

The late Stephen J. Gould, Harvard paleontologist extraordinaire, author, and baseball fanatic, argued in his book *Full House* that *reification* is the big danger that arises from averaging. According to Gould, the term was coined by philosophers and social scientists in the mid-nineteenth century and "refers to the mental conversion of a person or abstract concept into a thing. . . . We abstract the variation within a system into some measure of central tendency like the mean value and then make the mistake of reifying this abstraction and interpreting the mean as a concrete thing." In other words, we forget that the mean value of some natural trait is just that, the mean or the average—a value that may never actually occur on the ground.

Scaling up occurs when short-term observations or predictions are scaled up to make long-term predictions. The mother of all scale-ups is, of course, the modeling effort at Yucca Mountain. A database involving, at best, decades of observations is used to predict radioactive waste behavior a million years into the future.

Substituting laboratory measurements for nature is often necessary because observing the same relationships in nature is complicated by too much statistical noise, not to mention the time, expense, and difficulty of getting measurements in extreme events. But the lab is never as good as the real thing. Substitutes for the natural world include wave tanks and flumes to imitate surf zones and rivers, wind tunnels to study sand transport, greenhouses to observe plant response to climate, and aquariums to study aspects of fish development.

Substituting mathematics for nature can be even shakier than using lab measurements. The Army Corps of Engineers explained that it used a mathematically derived offshore profile shape (equations 4 and

5, appendix) to put into the model GENESIS because natural profiles on the North Carolina Outer Banks, of which many measurements were available, were too variable!

Another common problem, one that is very difficult to address in quantitative models, is the assumption of *linearity* or *nonlinearity*. In most modeling of complex systems, the relationship between parameters is assumed to be linear. The relationship between parameters forms a straight line on a graph, making it easy to perceive, easy to work with. In reality, simultaneous relationships between multiple parameters in models—for example the relationship between wave-formed currents and sand transport on beaches—are very complex and usually nonlinear. Such relationships do not plot as a straight line and become difficult, if not impossible, to handle in models.

It is an axiom of mathematical modeling of natural processes that only a fraction of the various events, large and small, that constitute the process are actually expressed in the equations. Note the list of factors that affect shoreline retreat rates in chapter 4 and the list of parameters responsible for sand transport on beaches in chapter 6. The hope, of course, is that the omitted processes matter little. Unfortunately, *omission of important processes* is commonplace. The failed model predictions attributed to so-called unusual events, such as an unusual storm, an abnormal flood, an unexpected wind, an extraordinary rainfall, unanticipated temperatures, or atypical water compositions, are often caused by an omitted process, which was incorrectly assumed to be unimportant. For example, the role of bacteria in chemical reactions in abandoned open-pit mine lakes is commonly omitted, as is the role of wind in determining the velocity of the surf zone currents that carry sand. The common "unusual event" excuses for failed modeled predictions, more often than not, are actually omitted processes.

The processes that transport sand on the seafloor near the shoreline provide a good comparison of the difference between the world of modeling and the real world. If you had just arrived on a spaceship from Mars and found yourself examining the world of coastal science, you could be forgiven for thinking that there must be two oceans. One is the ocean as envisioned by those who mathematically model it; the other is the real ocean. The number of factors in the real ocean that affect seafloor changes and sand transport is large, and the number of permutations and combinations of these factors is vast. The modeler's ocean, however, is slightly more complicated than a tub of water, with waves that are of perfectly uniform size, all coming from the same direction over a smooth, featureless seafloor covered by a blanket of perfectly uniform sand grains.

Initial conditions must be well known before an earth process or an earth system is modeled. *Lack of knowledge of initial conditions* can "effectively prohibit detailed prediction of system evolution," according to Haff. Dependence upon initial conditions is an important characteristic of chaotic behavior, as illustrated by the classic experiments of MIT professor Edward Lorenz in the early 1960s. Lorenz's famous butterfly effect, employed in every textbook about chaos, is about initial conditions. The story goes something like this: a butterfly flaps its wings in the Amazon, creating a very minor atmospheric disturbance that leads eventually, through many steps, to a tornado in Texas.

As explained by James Gleick in his book *Chaos*, "Tiny differences in input could quickly become overwhelming differences in output"—a phenomenon he describes as "sensitive dependence on initial conditions."

Gleick relies on folklore to provide an example of extreme sensitivity to initial conditions:

> For want of a nail, the shoe was lost
> For want of a shoe, the horse was lost
> For want of a horse, the rider was lost
> For want of a rider, the battle was lost
> For want of a battle, the kingdom was lost
> And all for the want of a horseshoe nail.

External forcing (forces intruding from the outside) occurs in so-called open systems where, according to Haff, "mass, energy and momentum can enter and be discharged through the system boundaries." He also notes that characterization of external forcing becomes an increasing problem for modeling as the size of the natural system increases. The most important form of external forcing in beach modeling is randomly occurring storms that pass right through or close enough to the shoreline in question to produce waves.

Storms are external forcing elements for invasive plant pests as well. Hurricanes frequently blow in new species (such as soybean rust, which may have arrived with Hurricane Floyd in 1999). Ocean currents are external forces for fishery modeling. Groundwater flow from outlying areas is an external forcing element for both Yucca Mountain and water quality quantitative modeling in general.

To Haff's list of model problems we would add ordering complexity, the problem we have discussed in most of the chapters. Even if all the parameters are thoroughly understood, the order and magnitude of their participation in the process remains unknown.

A Tainted Era

Among the early model skeptics, J. H. Chessire and A. J. Surrey in 1971 noted that because of the mathematical power of the computer, the predictions of computer models tended to become "imbued with a spurious accuracy transcending the assumptions on which they are based. Even if the modeler is aware of the limitations of the model and does not have a messianic faith in its predictions, the layman and the policymakers are usually incapable of challenging the computer predictions. . . . A dangerous situation may arise in which computation becomes a substitute for understanding a complex system." This, at a time when models were revered, sacrosanct, and seldom criticized.

If prediction by models is an albatross around the neck of those concerned with earth processes, it may be even more so for economists and others who model human behavior. Stock market trends and energy futures are as complex as natural processes, with the added complication of human behavior. William Sherden in his book *The Fortune Sellers* says, "So long as we do not question the validity of forecasts and think for ourselves, we will be destined to be deluged by a constant reign of error from those dismal scientists [economists] ever eager to fill our need for prediction."

In this volume we have viewed at least two dozen different kinds of quantitative model efforts, seven in particular detail. Most of these efforts to predict the outcome of complex natural and human related processes involve using several models in tandem. At Yucca Mountain, the modeling effort combines hundreds of models. We believe there are none that can predict accurately the outcome of complex natural processes, but there are degrees of differences in the recognition of model weakness among those that do the modeling.

Table 9.1 is a purely subjective ranking of the flavor of quantitative modeling effort in various science and engineering specialties. This is not a ranking of predictive modeling capabilities, since in all of these cases we believe actual accurate predictions are not a possibility, now or in the future. Instead, the ranking is based on the degree to which

Beach Nourishment Life Spans	WORST
Bruun Shoreline Erosion Rates	
Abandoned Mine Pit Water Quality	
Yucca Mountain Nuclear Repository	
Allowable Fish Catch	
Global Sea-Level Change	
Invasive Plants	BEST

uncertainties are recognized and publicized by the particular modeling community, the vigor of the debate about model validity in the technical literature, and the usefulness of the modeling effort, not in predictive successes but in advancing our knowledge of natural processes.

Each of the modeling groups or specialties has its own distinctive personality. The modeling of beaches for coastal engineering takes the cake as the worst of the bunch. There is not the slightest public recognition of problems with the models, and the models are so tightly bound up with politics that the answers are essentially useless, except to those whose purposes are served by inaccurate answers. Fudge factors are routinely used, looking back to learn from the past is simply not done, and most practitioners remain blissfully unaware of, or at least uncaring about, model weaknesses. The problem is amplified because engineers who use the models are rarely specialists in sedimentary geology, and they fail to appreciate the model's detachment from reality.

Engineering models for highways, buildings, and elevated water tanks usually afford some opportunity for correction, and unanticipated problems can be corrected during construction or initial testing. Such options don't exist for models of natural processes, however. Naomi Oreskes notes that "modern aeronautical designs are developed on computers [models] but no one ever buys a ticket on a commercial jet before a prototype has been flown for many hours." The problem is that engineers often don't recognize the difference between the behavior of natural processes and the behavior of steel and concrete.

Artificial beach life spans, shoreline erosion rates, and abandoned pit lake composition modeling are generally not adding significantly to our understanding of natural processes. The modeling efforts in these three fields follow separate and independent paths from current field investigations. The other modeling efforts listed in the table have spawned detailed studies that have contributed immensely to the science. Even if ordering complexity prevents accurate prediction of the future in these fields, the increase in knowledge of the causes of global change, fishery science, invasive plants, and the evolution of waste stored at Yucca Mountain will serve to provide a strong basis for qualitative estimates or risk analyses. The advance in knowledge of groundwater transport through rocks above the groundwater table (the vadose zone) has been a particularly fruitful aspect of the Yucca Mountain studies.

Agency incompetence and intransigence such as that exhibited by the Bureau of Land Management or the Army Corps of Engineers add another element to the spreading use of models. Without the models that provide favorable cost-benefit ratios and positive environmental impact statements, the agencies would be virtually out of business. In part this situation has been caused by the U.S. Congress, which requires some government entities, especially the Corps of Engineers, to sing for their supper. No projects—no budgets—no agency—no jobs. Under such circumstances it is no wonder that nonworking mathematical models are accepted unflinchingly if they will lead to project approval.

Policy scientist Ron Brunner thinks that part of the problem of model overemphasis is scientific hubris—if Newton and Einstein could do it, so can we. He thinks another driving force is the *law of the hammer*. To a small boy with a hammer, everything looks like a nail. To a scientist schooled in modeling, mathematical models answer all questions. Victor Baker, former president of the Geological Society of America, says, "Allowing the public to believe that a problem can be resolved . . . through elegantly formulated . . . models is the moral equivalent of a lie."

Totally consumed by the belief that to be quantitative is to be in the forefront of science, modelers consider nonbelievers to be neo-Luddites. Responses to criticisms and excuses for failed predictions follow several predictable lines, such as the following:

- The storm (flood, sea level rise, pit lake composition) was entirely unexpected due to very unusual conditions.
- We're learning from our mistakes.
- This is the best model we have at our current state of knowledge and until we find something better, don't throw the baby out with the bathwater.
- Models aren't all that important—we just use them to fine-tune results and as a check on other approaches and tools we use.
- Simply criticizing the assumptions behind the models is not enough. It is necessary to run the model and check it out.

Often modelers will note weak assumptions and sources of error in the technical literature but then pass right on by and present their model and recommend its application elsewhere. It's as though a difficulty can be overcome by recognizing that it exists. So it went with the Army Corps of Engineers technical manual on the SBEACH model. A tabulation of virtually devastating problems and uncertainties is scattered throughout the manual for model users, but in the end the model is recommended for use. As mentioned earlier, use of the model for nourished beach design is even required by law in Florida. Passing by weak assumptions is what Ron Brunner refers to as "uncertainty absorption."

Two more critical aspects of quantitative modeling remain to be mentioned. These are hindcasting and the demarcation problem otherwise known as the white swan problem.

Hindcasting is not forecasting, but it is widely accepted that reproducing the past is the same as a successful prediction of the future. This is one of the most common and misleading missteps of modern quantitative mathematical modeling, especially among academic scientists. For example, because a model successfully "predicts" past climate changes the assumption is made that it will predict future changes. Although this belief has been challenged a number of times (e.g., Naomi Oreskes), it still persists. But it is clear that in complex natural systems, successful prediction of one event doesn't mean that it will work the next time the model is applied.

Demarcation is the term used by science philosophers to describe the problem of distinguishing bad science from good science. Sir Karl Popper, an Austrian-born British philosopher of science, argued that a valid scientific theory must be based on falsifiability. He illustrated this by his famous tale of an individual who, convinced that all swans were white, looks only for white swans to verify his belief. Popper's point is that the person ought to be searching for black swans, since a single black swan would mean his intuition was wrong. So it is with models. Scientific mathematical modeling should involve constant efforts to falsify the model. To do otherwise is invalid science. It is easy to find evidence to support a model, a theory, or one's intuition by looking only for the "white swans." For example, the Bruun Rule model for predicting shoreline erosion rates was validated by finding a few locations where it seemed to work (the white swans) while ignoring many other locations where it didn't (the black swans). Opponents of the reality of future sea-level rise who are motivated by economics are constantly looking for white swans. Of course, the white swan problem affects all of science, as well as other segments of society, such as those who predict trends in stock market prices. But the impervious nature of the inner workings of models makes them particularly vulnerable to selective calibration or concentration on finding only white swans.

The Damage Done

That applied quantitative modeling has been damaging to our society is obvious. We have provided numerous examples of this in the previous chapters. Whether directly the result of using models to avoid data gathering, the politicization of models, the use of models as fig leaves, or the use of models by those who have no idea what they really are, the results are the same. Society is misled. Society loses.

A common justification for modeling is that it allows understanding of processes that are difficult or very costly to study in the field, such as the processes in a surf zone. With increasing frequency, however, much-needed field and laboratory studies are a casualty of modeling. Modeling offers a handy excuse to avoid investing in a lengthy and costly field study; why do so when one can just sit down at a computer and solve the problem?

Another victim of modeling can be robust science. It is the nature of scientists to be questioning and skeptical, a treasured tradition that has resulted in legendary societal debates on issues ranging from the origin of the universe to the evolution of life to the causes of disease. Our experience indicates that studies that are critical of quantitative models are most often completely ignored. In the technical literature, with the occasional exception of climate change modeling, the modelers circle the wagons to form a protective shield around their numbers. They are not unlike religious fanatics. Applied mathematical modeling has become a science that has advanced without the usual broad-based, vigorous debate, criticism, and constant attempts at falsification that characterize good science.

The widespread use of model simplification goes against the grain of good science even though it is a perfectly valid basis for qualitative modeling. A model simplification is an assumption or concept that is generally wrong but is believed to be close enough to reality to be used to make its application much simpler. Assuming, for example, that the highest one-third of all waves is a good measure of wave height has become a widely used simplification in equations used to calculate beach sand transport. But in a number of modeling communities, simplifications have drifted into scientific principles. For example, the study of beach sand transport has brought the concept of a shoreface profile of equilibrium related only to grain size and the offshore sediment fence known as the closure depth into mainstream science. The problem is that neither of those things exists in the real world.

Robust science is also damaged in the publishing process. The concept of the various categories of model uncertainties formulated by Peter Haff and discussed above is from a technical article in a rather obscure publication. It is a major contribution to the understanding of model weaknesses, but it was flatly turned down by the first (and much more prestigious) journal he submitted it to. Basically the editor and reviewers said that there was nothing new here; the uncertainties were well known by modelers. Policy scientist Ron Brunner had the same experience with a paper that was critical of global change modeling. He was informed that his paper was beating a dead horse—exactly the criticism that Haff's paper received. Brunner eventually published his paper, after three rounds of reviews. He accomplished this by carefully documenting that although modelers may have recognized the weaknesses as they had claimed, they had failed to mention this in the technical literature or in any public setting.

The integrity of science is an issue as well, especially in politically sensitive modeling. Unjustified claims of modeling successes abound. A frequent claim is that past model applications have been monitored to determine success or failure and to improve the models. This, however, is rarely the case when it comes to earth surface process modeling. Even if monitoring is done, true believers are never the best judges of their accomplishments and tend to monitor through rose-colored glasses. Claims of looking back or monitoring the results of modeling must be looked at very carefully and skeptically. It is critical to lift up the flap of the tent and take a close look at the underpinnings of such claims.

James Wilson, the University of Maine fishery economist, sums up his academician view of mathematical models in U.S. fisheries as follows:

> The models are not verifiable and no attempt is made to verify. The result is that we don't learn, except at an extremely slow rate. Almost all the science is done in government facilities and what's not done in government facilities is done on government contract. There is no independent body capable of giving depth and breadth to an alternative view. Academic scientists, including ecologists, often get involved but find that the National Marine

Fisheries Service becomes an attack machine when there is any substantive disagreement. In short, the institutions that keep science vibrant and progressive are absent.

And society suffers accordingly.

Societal damage from applied models can be extreme. After World War II, the RAND Corporation took over operational research for the U.S. Air Force and quickly changed its nature. It soon became systems analysis, which addressed the problem of what new equipment and tactics were needed to accomplish the mission of holding the Soviet Union at bay. During the war, operational research had examined existing tactics using existing equipment. Systems analysis, on the other hand, dealt with unknowns in the nearly complete absence of data. Operational research was concerned with military experience and was based on hard facts.

The RAND Corporation's quantitative model studies were behind many of the cold war decisions that eventually cost and perhaps wasted billions of dollars. Model-backed decisions included the Strategic Air Command's choice to build intercontinental bombers, as well as the decision to substantially increase the size of our nuclear stockpile. Among other things, RAND perceived a missile gap with the Soviet Union when our missiles were far better and much more numerous. Secretary of Defense Robert McNamara approved construction of hundreds of Minuteman missiles, bulldozing intelligence estimates and common sense, all the while shielded by the fig leaf of the RAND quantitative mathematical models.

Paul Edwards, in his 1995 book on computers and the cold war, argues that enormous military investments were based primarily on mathematical models that used assumptions that were based on the ideas and opinions of the civilian analysts at RAND. "The appearance of hard answers achieved by extensive quantitative analysis and simulation lent an air of certainty to results even when based on uncertain assumptions, especially at a moment in American history when the prestige of science and technology had reached an all time peak." In the end, the models just expressed some opinions.

A Qualitative World

Whatever alternatives society chooses to replace quantitative models, it will first be necessary to make fundamental changes in our approach to designing with nature. Accurate estimates of the outcome of natural processes must not be expected or required. Cost-benefit ratios must become a thing of the past, because determination of both costs and benefits requires accurate quantitative models. Accurate environmental impact predictions are impossible as well and should be considered ball-park figures. Accurate predictions of future climates, sea-level changes, shoreline erosion rates, and fish populations should be recognized as im-possibilities. We will simply have to move into a more qualitative world.

In many ways we are already there. William Gray, the hurricane guru of Colorado State University, predicted in December 2004 that the 2005 hurricane frequency would be above average (six hurricanes and eleven named storms) but not as bad as the 2004 season. In the year of Katrina, all bets were off, all predictions were wrong, and the naming of storms used up A through Z and dipped into the Greek alphabet.

The much-respected Gray utilizes a purely qualitative modeling ap-proach, hindcasting the relationship between past atmospheric condi-tions and hurricane activity. Previously he used the rainfall pattern in West Africa, but for reasons that are not clear, after 1995 this approach seemed to fail. The new approach uses six atmospheric predictors that seem to be related to hurricane activity. His is a *statistical model* that does not require an understanding of why a relationship exists—only ac-knowledgment *that* it exists. Gray says that the method is valid "provided the atmosphere continues to behave in the future as it has in the past. We have no reason for thinking that it will not."

To reiterate, qualitative models are those that answer what if, how, and why questions as opposed to the where, when, and how much ques-tions that quantitative models tackle. Quantitative models are expected to produce answers accurate enough to be useful for a wide range of societal purposes. Qualitative models are supposed to predict directions, orders of magnitudes, and the mechanisms behind natural processes.

"What if " qualitative modelers suppose that an event of some kind will occur and then evaluate what the consequences might be. As part of their study of the ongoing drought in Arizona, for example, geologists and hydrologists are concerned with understanding the dropping level of Lake Powell behind Glen Canyon Dam. The lake level has dropped 137 feet since 1999.The worst known drought in the area, examined through tree ring studies, occurred in the 1500s and lasted for thirty-eight years. Hydrologist Ron Harding asked the question What if we assume a drought that is drier but lasts as long as the sixteenth-century drought? This was a what-if worst-case scenario. Harding modeled the hypotheti-cal plummeting lake level, with the startling result that the lake level now is plummeting faster than it does in the modeled, worst-case natural scenario. The reason, of course, is the heavy use of Colorado River water by Nevada, Arizona, and California, superimposed on the lake water lowering caused by the drought. This what-if qualitative modeling pointed out the gravity of the lake level drop.

The dikes that failed in Hurricane Katrina around New Orleans provide a tragic what-if qualitative modeling case in point (figure 9.1). Louisiana State University modelers predicted years ago that the Mississippi River Gulf Outlet (MRGO) canal would accentuate storm surges and endanger New Orleans. The model-based warning was ignored, and sure enough, the canal raised the storm surge at the head of the canal, causing a dike failure at that point.

An excellent example of a qualitative model used to answer a "how" question involves a study of the origin of tors, which are rock pillars or crags a few tens of meters high on the surface of pediments. Pediments are broad aprons at the base of many mountain ranges in the arid southwestern United States, for example, the Wonderland of Rock in Joshua Tree National Park. They are large flat erosion surfaces, covered by thin layers of sand and gravel.

Why and how tors form has been the source of much discussion among geologists. Current theories mostly center on the idea that there are variations in the underlying rock types, and the fact that some weather faster than others may explain why the rock knobs are present. That is, the tors would be the most resistant rock. Alternatively, some argue that fractures and faults may cause localized accelerated weathering, thus isolating adjacent highs where weathering is slower.

However, many tors form on pediments where both the frequency of fractures and the rock types are essentially uniform, so the prevailing theories can't always explain them. Mark Strudley, a modeler and graduate student, theorized that tors might form because unconsolidated sediment surfaces (and the underlying rock) are gradually lowered at a faster rate than bare rock alone. Such rates of lowering are in the range of tens of meters per million years. By this hypothesis, the tors form where conditions favor a very thin to nonexistent sediment cover. Strudley modeled pediments and the processes of evolution on its surface. The model produced a tor shape and an areal distribution of tors (on paper) that was similar to actual tor distribution (figure 9.2).

It would have been possible to theorize the relationship between sediment thickness and tor evolution (especially with extensive fieldwork) without the model, but the use of the qualitative model provided

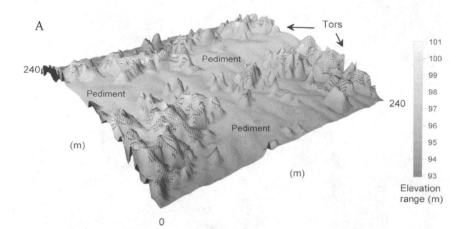

A

Tors

Pediment

240

Pediment

Pediment

(m)

240

(m)

101
100
99
98
97
96
95
94
93
Elevation
range (m)

0

B

Figure 9.1 An example of the use of a qualitative mathematical model. Here the question was how do tors, the rocky crags that protrude from the desert surface, form? Modeler Mark Strudley assumed that tors forms where relatively thick sediment covers retarded weathering, allowing adjacent surfaces with thinner sediment cover to lower more quickly. The resulting modeled image of the desert surface (A) is remarkably like the real thing (B). This shows that Strudley's proposed mode of formation of tors is feasible, but it does not prove that this is the mechanism that is actually in play. Photos courtesy of Mark Strudley.

Figure 9.2 The flooding caused by Hurricane Katrina could hardly be considered unexpected. This photo from the downtown area of New Orleans shows relatively shallow but still very damaging flooding. Qualitative modeling showed long ago that the flooding was likely when the "right" storm from the "right" direction came by. One model correctly predicted that the Mississippi River Gulf Outlet (MRGO) navigation channel leading from the city to the open Gulf of Mexico would act like a funnel and enhance the level of the storm surge, causing the water to overtop the dikes. Photo courtesy of the U.S. Army.

strong support for the idea. In addition, the computer model allowed examination of surface behavior that would not have been possible to derive with simple reasoning of the human mind. The fact that the model reproduced reality, however, is not absolute proof that this mechanism is responsible for tor formation. Also, of course, the other mechanisms theorized by earlier workers may well be operable at the same time.

Probably the main opposition to casting aside the quantitative approach to prediction of the path of natural processes will be the engineering community, for they are the ones who brought mathematical modeling out of the world of concrete, asphalt, and steel, where the laws of physics prevail, into the chaos and complexity of the physical and biological processes that work on the earth's surface. Quantitative engineering modeling is a very successful component of modern engineering practice. Engineers excel at predicting the impact of natural processes on human-made structures, such as the effects of wind on a building. That is a different problem from predicting the outcome of natural processes such as storm winds on a beach or the eventual composition of the lake in an abandoned open-pit mine.

In our personal interactions with engineers we have learned that change may not come easily. A few years back we presented a paper arguing that the model GENESIS, which predicted the evolution of artificial beaches, could not possibly work. Afterward an engineer came up and expressed outrage concerning our study. After listening to her tirade for a while, it became clear that she was not questioning the veracity of our findings. She was offended that we had dared to question such an important model.

In a recent question-and-answer session, again concerning beach models, an engineer from the U.S. Army Corps of Engineers responded to criticisms by using all the standard answers, such as the assertion that the models had been calibrated and verified and there were no problems with using average grain size and average wave height. When it was pointed out that his model did not consider storms, he noted that his organization was in the process of correcting that. Although he was basically wrong on all counts, what soon became clear was that he was as certain of the model's validity as he would be of the formulas used to determine the area of a rectangle and the circumference of a circle. That it might be wrong was not a possibility.

Both engineers in these two encounters are *grunt engineers*—the ones "in the foxholes" who actually apply the models. They are not research engineers, trained to probe and question. They have been trained to apply models unquestioningly in the rote fashion characteristic of the undergraduate curriculum in some engineering schools.

One fascinating example of engineering modeling of the impact of nature on human-made structures was the problem of fractures in World War II Liberty ships. Of the 2,700 vessels manufactured, 400 of them experienced hull fractures, 90 of which were serious and perhaps 20 of which resulted in the ship's sinking. All fracturing occurred in the cold water of the North Atlantic; the phenomenon was virtually nonexistent in ships traversing warm South Pacific waters. After a Liberty ship broke completely in two in 1943 while traveling between Siberia and Alaska and especially after another broke in two while docked in Boston Harbor in the winter of 1947, study of the phenomenon became a high priority. Before the Liberty ship problems, fatigue of metal plates was recognized, but low-temperature brittle fracture was not.

Eventually, mathematical modeling studies of the effect of the cold-water environment on metal plates identified the basis of the ship-fracture problem and a new field, *fracture mechanics*, was born. In hindsight, one might speculate about whether cold-water fracturing could have played a role in the sinking of the *Titanic*.

A World Without Models

Adaptive management has been discussed in several of the preceding chapters, including application in managing the Yucca Mountain nuclear repository and our disappearing fisheries. In the management of marine reserves for commercial fisheries, adaptive management involves trial and error rather than reliance on accurate predictions to determine the size and location of no-fish zones. The idea is to adjust the no-fish zones according to the success or failure of the initial reserve designation. If the sought-after increase in BOFFFs (big old fat female fish) is not achieved at first, the approach is altered. *Adaptive staging* or *adaptive management* could undoubtedly replace the quantitative predictive approach in many areas.

One huge area of quantitative mathematical model application is energy (coal, oil, nuclear). including availability, limits, costs, relative feasibility of solar, hydro, wind, nuclear and fossil fuel, energy and the environment, energy and war, and much more. Human behavior plays a big role here. For example, in the early 1970s Glenn Seaborg, chairman of the Atomic Energy Commission, projected a nuclear-electrical-generating capacity of 2,100 million kilowatts by the year 2000. The reality that came to be was 780 million kilowatts—the models didn't know that people would begin to object to nuclear power. Another model foiled by human behavior!

Vaclav Smil, a University of Manitoba geography professor and energy expert, argues that in energy affairs, model failures are a way of life. "The dismal record of long range forecasts ... demonstrates convincingly that we should abandon all detailed quantitative point forecasts." The models used to forecast energy trends are referred to as *consolidative models*, or models in which the facts are brought together into a very complex model that is supposed to imitate the real world. (Each modeling specialty seems to have an independent lexicon of terms to describe its models. Use of the term *consolidative models* is clear evidence of the need to consolidate model terminology!) Smil cautions, however, that consolidative modeling doesn't work if there is "insufficient understanding" or if there are "irreducible uncertainties." The global energy system that Smil studies certainly falls into both of these categories, as do fisheries, groundwater flow, global climate change, beaches, and pit lake models. Smil believes, however, that modeling is here to stay and that we will continue to spend lots of effort and money on prediction but will never get better at it. "We will not do better as we try to include every conceivable factor in our assessments because many are either unquantifiable or their quantification cannot go beyond educated guessing."

Smil believes that a small calculator and the back of an envelope often can provide answers as useful as those obtained from modeling forecasts. He suggests that most of the grossly inaccurate forecasts of energy trends can be explained by two human truths. One is the mood of the moment, the tendency to be strongly influenced by current events and recent trends. The second is a fixation on new technology, preferred policies, and simple magical solutions.

He suggests that there are two non-modeling means of looking ahead that could be much more useful to society: *contingency scenarios* and *normative scenarios*.

In the contingency scenario approach, various possible scenarios are considered, ranging from those that are currently deemed likely to those viewed as most extreme. In the world of energy forecasting, extreme scenarios might include a global economic depression, general war in the Middle East, or terrorist attacks on pipelines. In the case of Yucca Mountain, the extreme scenarios would be a nearby volcanic eruption, a major earthquake, or a change in climate from arid desert conditions to a tropical rain forest. With this technique there is hope that no matter what scenario comes to pass in the future, an appropriate and rapid response has already been planned.

Smil suggests that society could determine norms of behavior, or normative scenarios, and action "to guide our long term paths toward the reconciliation of human aspirations with biospheric imperatives." This is an approach where humans strive to live in harmony with nature. Society determines what should happen and works in that direction rather than trying to predict what will happen and then riding along with events. When goals for a better life and a better environment are determined, society should then work toward achieving them.

Table 9.2 Some Distinguishing Characteristics of Scenario and
Strategic Planning

SCENARIO PLANNING	STRATEGIC PLANNING OR MATHEMATICAL MODELING
qualitative input	quantitative input
exploits uncertainties	minimizes uncertainties
long-range planning	short-term planning
multiple answers	single answer
planning the future	predicting the future
hypothetical events	predetermined goals

The failed Kyoto Treaty, intended to globally reduce carbon dioxide emissions, was an example of the normative scenario approach. The same approach in the case of Yucca Mountain, given that we must have a repository for society's safety and well-being, would be the purchase of the property where radioactive waste might flow in case of a disaster and the moving of people off the property and out of harm's way.

University of Texas law professor Philip Bobbitt refers to the contingency scenario approach as *scenario planning*, "the construction of alternative scenarios rather than single point predictions in order not so much to predict the future as to help policy makers think about the future." Royal Dutch Shell greatly improved its fortunes by adopting scenario planning in the early 1970s. Among various scenarios the company considered was the rise of OPEC. The modeling approaches we have discussed in previous chapters are virtually all single-point predictions. Bobbitt refers to these as *strategic planning*. Table 9.2 lists some distinguishing characteristics of scenario and strategic planning.

Surely scenario planning (or the contingency scenario approach) holds promise as a replacement for the modeling approach currently practiced. If the management of the cod fishery had been based on scenario planning (and if politics hadn't intervened), the size and nature of both the nearshore and the offshore catch, as well as the abundance of the cod's food supply, the size of recruitment classes, and other factors could have all been incorporated into possible scenarios, and plans could have been laid for various "good" and "bad" events. Having done no such brainstorming, however, and depending solely on the size of the catch as the basis for management, mixed with a heavy dose of politics, the Canadian government allowed the world's greatest fishery to collapse. Other fisheries seem to be following this same disastrous path. Needless to say, it is possible for politics and special interests to spoil the best scenario planning.

What if contingency planning had been in place for abandonment of the Berkeley Pit in Butte, Montana? Would the state and the community have decided to prevent the world's largest cup of poison from forming by continuing to pump the pit dry? Would the decision have been to continuously decontaminate the water as the pit filled?

Contingency planning can be broadly applied to the use of quantitative mathematical models. For example, one (very likely) scenario is that the prediction by the model will be wrong. Planning for contingencies—if the pit water is more acidic than predicted, if the artificial beach disappears faster than assumed, if the fish numbers continue to decline, or if sea level rises faster than predicted—will place response plans in the hands of the responsible officials.

Quantitative applied models of processes on the surface of the earth for practical applications in engineering, policy, and environmental management should go the way of the passenger pigeon. Unfortunately, however, it is a fair prediction that applied quantitative modeling of complex systems will continue and even accelerate in our society, at least until public skepticism and recognition of failures arrest the trend. For those who understand the absurdity of quantitative applied models, some version of scenario planning and adaptive management may be the right path.

Thinking Like Physicists

Some energy companies like to hire distinguished physicists to devise mathematical models predicting future energy trends. Presumably this employment opportunity comes about because physicists are assumed to be applied mathematical whizzes, which they usually are. Physicists, however, are the last ones the industry should hire. Better to sign up scientists, like geologists, with a quantitative bent who deal with earth processes.

Physicists generally deal with a noncomplex world governed by rules and laws that are hard and fast, such as the principle of a ball rolling down an inclined plane, the constant rate of radioactive decay of uranium, and the predictable orbits of a planet around a sun. Once the physics of a ball rolling down an inclined plane is understood, prediction of the ball's velocity on planes set at various angles can be accurately and transparently accomplished. All who apply the laws of physics to this experiment will arrive at the same answers.

But complexity rules energy consumption, fuel prices, and coal production rates, and such complex energy systems cannot be quantitatively modeled. The mathematical models of different experts will come up with radically different predictions, as is apparent almost daily in TV business reports about the price of gasoline, heating-oil futures, and the Middle East petroleum outlook. The same is true for earth surface processes. The prediction of the future of beaches, fish abundance, the level of the sea, rates of shoreline erosion, and the future of invasive plants is entirely separate from the ordered, predictable, comfortable world of physicists.

Thinking like physicists and not recognizing complexity is what has allowed us to escape from reality through quantitative mathematical modeling. It has allowed us to predict the unpredictable. Ironically, in the unwritten pecking order of the sciences, physics and math are number one because they are the most quantitative, and geology bottoms out because it is dealing with many aspects of earth surface processes that cannot be quantified. The arrogance of some physicists who share this view is legendary. Ernest Rutherford, the father of nuclear physics, is credited with the observation that "all science is either physics or stamp collecting" (more recently, Robert McNamara referred to the qualitative approach as "poetry"). Unfortunately, as a result of this meaningless hierarchy, physics envy is a factor driving the rush to prediction by mathematics. If it is quantitative, like physics, then it has got to be sophisticated, like physics!

The quantitative trumping the qualitative has been with us since Lord Kelvin's time. This is what happened when the American Weather Service ignored qualitative hurricane warnings by the Cubans, who were reading cloud patterns instead of using modern weather-forecasting techniques. The result was that no one left Galveston before the 1900 hurricane struck and 6,000 souls perished.

Geology and biology are sciences that depend highly on field observations and expert intuition. Quantifying the behavior of organisms, the oceans, hill slopes, or sand grains with models requires stepping out of the intricate, dynamic, and supple world of nature into a wooden, unyielding, and inflexible world. It is a world where mathematical equations characterize events and processes, equations that can describe only a small part of the picture in very simple fashion. The intuition of an experienced scientist is gone. At best, only a small fraction of the processes that lead to the desired endpoint prediction can be considered.

The widespread use of coefficients (in the beach behavior models, for example), which in reality are fudge factors, and the very common and usually invisible practice of tuning or tweaking models to come up with the right answer, open the door to political pollution. Applied models, out there in the midst of politically sensitive societal issues, are easily moldable to favor a cause, be it a cost-benefit ratio, an environmental impact statement, the cause of a disease, the size of a fish population, or a prediction of future hazard potential. Add a fudge factor here and tweak the model there, and you have the "correct" answer. And the alterations are invisible to the managers who use the models.

The "transparency and openness to evaluation" required for invasive plant models by the National Academy of Sciences panel is rarely seen in the modeling literature. It is ironic that invasive plant biologists, in a specialty that does not depend on models, are the group suggesting the strongest restraints and highest standards for model use.

Qualitative modeling to help understand such processes has a bright and productive future. Here the intuition of scientists is not discarded. Experience is used for, among other things, figuring out what parameters in the process that's being modeled can be ignored. With a qualitative approach, these models can sort out the most important causes of shoreline erosion, help explain why sea level is rising, and speculate about the effectiveness of reduction of carbon dioxide emissions on global warming. But no models can predict with useful accuracy the rates of shoreline erosion, the rates of sea-level rise, or the impact of CO_2 reduction.

This means that we have to abandon our claims of accurate prediction of many natural processes. Instead of accepting a prediction of a two-foot rise in sea level over the next century, we might note that sea-level rise is likely to continue and that its acceleration is a good possibility. Instead of claiming that wastes from Yucca Mountain nuclear waste repository will not escape for 10,000 or 1 million years, the public might be informed that escape of the waste is unlikely in the next few hundred years, but when it does escape it will flow in this direction, to this location, and create the following problems. Rather than predicting that an artificial beach will last 8 years before more sand needs to be pumped onto it, engineers should announce that the artificial beach could last 6 years but also could disappear next week if the perfect storm occurs.

Differing degrees of societal damage can potentially be created by poor modeled predictions. The consequences of depending upon a model to correctly predict the configuration of new sandbars on the Grand

Canyon floor after a water release from Glen Canyon Dam are minute compared to the consequences of predicting catch levels for the Grand Banks cod fishery that employed 40,000 people. The consequences of failing to predict the life span of a nourished beach are small relative to under- or overestimating global sea-level rise.

Regardless of the societal importance of the modeling effort, if we wish to stay within the bounds of reality we must look to a more qualitative future, a future where there will be no certain answers to many of the important questions we have about the future of human interactions with the earth.

Even in the unlikely event that sometime in the remote future we will understand each of the numerous parameters and their interactions and feedbacks that control the event we are predicting, we shall never accurately predict the future. No one knows in what order, for what duration, from what direction, or with what intensity the various events that affect a process will unfold. No one can ever know.

"We all want progress," notes C. S. Lewis in his book *Mere Christianity*. "But progress means getting nearer to the place you want to be. And if you have taken a wrong turning [*sic*], then to go forward does not get you any nearer. If you are on the wrong road, progress means doing an about-turn and walking back to the right road; and in that case the man who turns back soonest is the most progressive man."

appendix

We have not included a single mathematical model in the text of this book. This is because we believe our conclusions can be made solidly without mathematics. Furthermore, we are writing for nonspecialists and nonmathematicians who we suspect would be repelled by differential equations and the likes (and wouldn't buy our book). It has been our experience that many people consider mathematical models to be opaque and impenetrable because the mathematics is beyond them. In our estimation this impassable nature of mathematics has allowed modelers to carry their trade far beyond the limits of reality, to the great detriment of our society.

Robert Moran, the geochemist concerned with pit lake quality, and Jim O'Malley, fishing industry representative, both view the mathematical modeling community as an unassailable and untouchable priesthood. We agree with this view. Because it is a priesthood, mathematical modeling has become a science that has advanced without the usual broad-based vigorous debate and criticism that characterize good science.

Critically reviewing models by examining assumptions rather than the math can bring a gale of fresh air into the modeling community. They need it. Criticism can only strengthen their specialty.

Here, for the benefit of the reader who may wish to further delve into the world of models, we present the mathematics behind three mathematical models that are used to predict some aspect of a societally important natural process. The models are critically discussed in the pertinent chapters, and brief descriptions of model weaknesses are included here as well. We begin with the very simple Bruun Rule (chapter 5) and progress through increasingly complex models, CERC and GENESIS (both discussed in chapter 6).

The Bruun Rule

This simple model is intended to predict the amount of shoreline erosion that will occur as a result of sea-level rise. It is virtually the only model that is supposed to accomplish this.

$$(1) \qquad R = \frac{L}{(B+h)} S$$

In actual application the terms in the equation cancel out to become:

$$(2) \qquad R = \frac{1}{\tan\theta} S$$

Equation 2 basically says that shoreline erosion is proportional to the slope of the shoreface. This being the case, all that is needed to determine the impact of sea-level rise is a navigation chart on which to measure the slope of the shoreface plus knowledge of the local sea-level-rise rate. The slope of the shoreface is sometimes determined by equations 4 and 5, on the assumption that slope is controlled by grain size.

Where:
R = shoreline retreat due to sea-level rise
S = sea-level rise
B = berm (upper beach and dune) height
θ = angle of the surface of the shoreface
h = depth at the base of the shoreface
L = width of the shoreface

Problems: The assumption that shoreface profiles are controlled exclusively by grain size of the sand is wrong. The assumption that the shoreface profile will remain constant as the sea level rises is unproven and highly unlikely. There is no evidence that slope of the shoreface controls shoreline erosion. The model can be applied only to sandy shorefaces, devoid of rocks, mud, and gravel and of uniform grain size throughout, a rare situation.

Our view of the Bruun Rule can be summarized by the poem below. We acknowledge inspiration for this questionable attempt at poetry from P. A. Larkin's 1977 poem "MSY," about the demise of the concept of the maximum sustainable yield for fisheries (chapter 1).

> The Bruun Rule
> 1954–2006
>
> Here lies the Bruun Rule, once tried and so true
> The only way to predict what a sea rise will do.
>
> It told us how long the beach condo will last
> And to know just when the beach will move past.
>
> It once was a pure model quite novel and fine
> But, alas, as always comes progress with time.
>
> With no link to reality it just doesn't work
> Yet on many of the world's beaches the model still lurks.
>
> The time has now come to be objective and fair
> Time to send these modelers back to their lairs.
>
> And halt the damage the Rule still can cause
> And from now on follow only nature's laws.
>
> The day has now come to go the qualitative way
> And in the boneyard of ideas let the old Bruun Rule lay.

Bruun, P. 1962. Sea level rise as a cause of shore erosion. Proceedings of the American Society of Civil Engineers. *Journal of the Waterways and Harbors Division* 88:117–130.

Cooper, J. A. G., and O. H. Pilkey. 2004. Sea level rise and shoreline retreat: Time to abandon the Bruun Rule. *Global and Planetary Change* 43:157–171.

Pilkey, O. H., and T. W. Davis. 1987. An analysis of coastal recession models: North Carolina coast. In D. Nummedal, O. H. Pilkey, and J. D. Howard, eds.,

Sea Level Fluctuation and Coastal Evolution, 59–68. SEPM Special Publication No. 41. Tulsa: Society of Economic Paleontologists and Mineralogists.

The CERC Equation

This equation is widely used to determine the rate of longshore transport of sand on beaches. Basically the model assumes that the energy of breaking waves plus the angle at which they strike the shore provides the momentum to transport sand laterally down the beach. The higher the waves, the stronger the potential current.

(3)
$$Q = k \frac{\rho H_b^2 \sqrt{g d_b}}{16 (\rho_s - \rho) a'} 2 \sin \alpha_b$$

Where:
Q = quantity of sand that is moved
k = sediment transport coefficient
H = height of the breaking waves
ρ_s = density of quartz sand
ρ = density of seawater
b = subscript indicating breaking wave conditions
g = acceleration due to gravity
α = the angle of the wave approach to the shoreline

Problems: Basic assumptions and model simplifications are far out of date. For example, the assumptions are made that only waves (not currents) move sand, that shorefaces are made of sand of uniform grain size, and that no rock or mud layers are present. The sediment transport coefficient (k) is either a fudge factor designed to be used in order to come up with a "reasonable" answer or a number based on a fair-weather study of a beach in Southern California. Only a small fraction of the actual parameters that control longshore transport are included in the model (see list in chapter 6). Lack of knowledge of when, where, and at what magnitude the many parameters that affect this process will be involved (e.g., when will the next storm strike?) (ordering complexity) is a fatal blow to accurate prediction by this model.

Cooper, I. A. G., and O. H. Pilkey. 2004. Longshore drift: Trapped in an expected universe. *Journal of Sedimentary Research* 74:599–606.

Komar, P. D. 1976. *Beach Processes and Sedimentation*. Princeton, N.J.: Prentice Hall.

USACE. 1984. *Shore Protection Manual*. 3 vols. Vicksburg, Miss.: U.S. Army Corps of Engineers Coastal Engineering Center.

Generalized Model for Simulating Shoreline Change (GENESIS)

This model is intended to determine changes in shoreline position (erosion or accretion) on shorelines with coastal engineering structures including groins, jetties, breakwaters, and seawalls. In a typical application it involves four separate models that (1) determine the slope of the shoreface, (2) determine sand loss around and through engineering structures, (3) determine longshore transport sand volume, and (4) determine the expected rate of shoreline retreat.

Slope of the shoreface

$$h = Ay^{0.67} \tag{4}$$

$$A = 0.067w^{0.44} \tag{5}$$

Where:

w = settling velocity of the sand grains in a water-filled tube (a method of determining grain size)

h = water depth

y = the distance offshore

A = scaling parameter related to grain size of the sediment

The slope of the shoreface is calculated to use in the longshore transport equation. The shoreface is assumed to be controlled by the grain size of the sand.

Sand loss around and through engineering structures

$$BYP = 1 - \frac{D_g}{D_{LT}} \tag{6}$$

$$PERM \tag{7}$$

Where:

D_g = maximum water depth at the tip on an engineering structure such as a jetty or a groin

D_{LT} = maximum water depth of longshore transport

BYP = the amount of sand that bypasses engineering structures

$PERM$ = sand permeability of engineering structures

The equation (6) determines how much sand is bypassed past engineering structures by longshore currents. *PERM* stands for permeability of the engineering structure, that is how much sand is lost through the interstices of the structure. *PERM* is based on user judgment. o is no sand transport and 1 would be a completely "transparent" structure.

Volume of longshore transport

$$(8) \quad Q = (H^2 C_g)_b \left[a_1 \sin 2\theta_{bs} - a_2 \cos\theta_{bs} \frac{\partial H}{\partial x} \right]_b$$

$$(9) \quad a_1 = \frac{K_1}{16(\rho_s/\rho - 1)(1 - \rho)(1.416)^{5/6}}$$

$$(10) \quad a_2 = \frac{K_2}{8(\rho_s/\rho - 1)(1 - \rho)\tan\beta(1.416)^{7/2}}$$

Where:

Q = amount of sand transported

ρ_s = density of sand

ρ = density of seawater

p = porosity of sand on the beach/shoreface

K_1 = empirical coefficient

K_2 = empirical coefficient

b = subscript indicating breaking wave conditions

θ_b = the angle of breaking waves to the shoreline

a_1 = nondimensional parameter (equation 9)

a_2 = nondimensional parameter (equation 10)

W = wave height

C_g = wave group speed

x = distance along shore

$\tan \beta$ = average nearshore bottom slope

This is the equation to determine the volume of sand moved by the waves on the beach (longshore transport)

Expected rate of shoreline change

(II) $$\frac{\partial y}{\partial t} + \frac{1}{(D_b + D_c)} \left[\frac{\partial Q}{\partial t} - q \right] = 0$$

$\frac{\partial y}{\partial t}$ = rate of change of shoreline position

D_b = maximum water depth of breaking waves

D_c = closure depth

x = distance alongshore

y- = distance perpendicular to the shoreline

t = time

q = source of sand or loss of sand other than by longshore transport

Problems: GENESIS has all of the problems of the Bruun Rule and the CERC equation. Whereas the CERC equation commits the sin of omission (too few variables), GENESIS commits the sin of commission (too many variables). Some variables are completely unknown, such as the permeability of structures and q, the loss or gain of sand in a seaward or landward direction. The authors of this model admit that adequate data for running the model are seldom, if ever, available. Frequently, averaged values are used, smoothing over great potential variability (wave height and direction, storms, profile shape).

Hanson, H. 1989. GENESIS: A generalized shoreline change numerical model. *Journal of Coastal Research* 5:1–27.

Hanson, H., and N. C. Krause. 1989. *GENESIS: Generalized Model for Simulating Shoreline Change*. Technical Report 89–19. Vicksburg, Miss.: U.S. Army Corps of Engineers Waterways Experiment Station, Coastal Engineering Research Center.

Young, R. S., O. H. Pilkey, D. M. Bush, and E. R. Thieler. 1995. A discussion of the Generalized Model for Simulating Shoreline Change (GENESIS). *Journal of Coastal Research* 11:875–886.

references

1. Mathematical Fishing

Beverton, R., and S. Holt. 1956. *The Theory of Fishing in Sea Fisheries*. London: Edward Arnold.

——. 1957. *On the Dynamics of Exploited Fish Populations*. Fisheries Investigations Series, vol. 19. London: Fisheries and Food Department, Ministry of Agriculture.

Casey, Jill, and Ransom A. Myers. 1998. Near extinction of a large widely distributed fish. *Science* 281:690–692.

Chantraime, Pol. 1993. *The Last Cod Fish: Life and Death of the Newfoundland Way of Life*. Toronto: Robert Davies Publishing.

Chipello, Christopher. 2004. Nothing but net: A fishing industry fades, as does a way of life in Newfoundland. *Wall Street Journal*. May 19.

Clover, Charles. 2004. *The End of the Line: How Overfishing Is Changing the World and What We Eat*. London: Ebury Press.

Department of Fisheries and Oceans. 1993. *Charting a New Course: Towards the Fishery of the Future*. Ottawa: Communications Directorate, Fisheries and Oceans.

Eagle, Josh, Sarah Newkirk, and Barton Thompson. 2003. *Taking Stock of the Regional Fishery Management Councils.* Pew Ocean Science Series. Pew Charitable Trust.

Kurlansky, Mark. 1997. *Cod: A Biography of the Fish that Changed the World.* New York: Walker.

MacKenzie, Deborah. 1995. The cod that disappeared. *New Scientist.* September 16.

McGoodwin, James R. 1990. *Crisis in the World's Fisheries.* Stanford, Calif.: Stanford University Press.

National Research Council. 1998. *Improving Fish Stock Assessments.* Washington, D.C.: National Academies Press.

Nickerson, Colin. 1998. Newfoundland farewell. *Boston Globe.* September 20.

O'Brien, Stephen. 1992. Cold War: The EC returns a volley in fish fight. *Ottawa Citizen.* April 10.

Pauly, Daniel, and Jay Maclean. 2003. *In a Perfect Ocean: The State of Fisheries and Ecosystems in the North Atlantic Ocean.* Washington, D.C.: Island Press.

Wallace, Richard K., William Hosking, and Stephen T. Szedlmayer. 1994. *Fisheries Management for Fishermen.* Auburn University Sea Grant Extension, MASGP-94–012. Auburn, Ala.: Auburn University.

Wilson, J.A. 2002. Scientific uncertainty, complex systems, and the design of common pool institutions. In T. Ostrom, ed., *Drama of the Commons.* Washington, D.C.: National Academies Press.

2. Mathematical Models

Baker, V. R. 1994. Geomorphological understanding of floods. *Geomorphology* 10:139–156.

Bernstein, P. L. 1996. *Against the Gods: The Remarkable Story of Risk.* New York: John Wiley.

Dowd, Kevin. 1999. Too big to fail. Cato Institute Briefing Paper No. 52. September 23.

Edwards, P. N. 1996. *Closed World: Computers and the Politics of Discourse in Cold War America.* Cambridge, Mass.: MIT Press.

Gleick, J. 1987. *Chaos: Making a New Science.* London: Penguin.

Gould, S. J. 1996. *Full House: The Spread of Excellence from Plato to Darwin.* New York: Harmony.

Haff, P. K. 1996. Limitations on predictive modeling in geomorphology. In B. L. Rhoads and C. E. Thorn, eds., *The Scientific Nature of Geomorphology,* 337–358. Proceedings of the 27th Binghamton Symposium in Geomorphology. New York: John Wiley.

Konikow, L. J., and J. Bredehoeft. 1992. Ground-water models cannot be validated. *Advances in Water Resources* 15:75–83.

Kurtz, Howard. 2000. *The Fortune Tellers: Inside Wall Street's Game of Money, Media, and Manipulation.* New York: Simon and Schuster.

Lowenstein, R. 2000. *When Genius Failed: The Rise and Fall of Long-Term Capital Management.* New York: Random House.

Oreskes, Naomi. 1998. Evaluation (not validation) of quantitative models. *Environmental Health Perspectives* 106:1453–1460.

Oreskes, N., K. Shrader-Frechette, and K. Belitz. 1994. Verification, validation, and confirmation of numerical models in the earth sciences. *Science* 263:641–646.

Repcheck, J. 2003. *The Man Who Found Time: James Hutton and the Discovery of the Earth's Antiquity.* Cambridge, Mass.: Perseus Book Group.

Sarewitz, D., R. A. Pielke, and R. Byerly, eds. 2000. *Prediction, Science, Decision Making, and the Future of Nature.* Washington, D.C.: Island Press.

Sherden, W. A. 1998. *The Fortune Sellers.* New York: John Wiley.

Smil, V. 2003. *Energy at the Crossroads: Global Perspectives and Uncertainties.* Cambridge, Mass.: MIT Press.

3. Yucca Mountain

Ashley, S. 2002. Divide and vitrify: Partitioning nuclear waste saves space, but it isn't easy. *Scientific American.* May 13.

Bodvarsson, G. S., C. K. Ho, and B. A. Robinson, eds. 2003. Yucca Mountain project. *Journal of Contaminant Hydrology* 62 and 63. Special issues.

Duncan, D. E. 2004. Do or die at Yucca Mountain. *Wired.* November.

Dyer, Russ, A. Van Luik, R. Linden, and R. Salness. 2002. In search of water: An update on Yucca Mountain studies. *GeoTimes.* March.

Goldberg, Jonah. 2002. Dead and buried: The crazy debate over Yucca Mountain and nuclear waste. *National Review.* April 8.

Macfarlane, Allison. 2000. Standoff at Yucca Mountain: High-level nuclear waste in the United States. In J. S. Schneiderman, ed., *The Earth Around Us: Maintaining a Livable Planet,* 283–299. New York: W. H. Freeman.

Metlay, Daniel. 2000. From tin roof to torn wet blanket: Predicting and observing groundwater movement at a proposed nuclear waste site. In D. Sarewitz, R. A Pielke, and R. Byerly, eds., *Prediction, Science, Decision Making, and the Future of Nature,* 199–230. Washington, D.C.: Island Press.

Nadis, Steve. 2003. Man against a mountain. *Scientific American.* March.

National Research Council. 2003. *Disposition of High-Level Waste and Spent Nuclear Fuel: The Continuing Societal and Technical Challenges.* Washington, D.C.: National Academies Press.

Porter, R. C. 2002. *The Economics of Waste.* Washington, D.C.: Resources for the Future.

Wald, Matthew. 2001. Radioactive waste site: A shift in strategy. *New York Times.* July 31.

Whipple, Chris. 1996. Can nuclear waste be stored safely at Yucca Mountain? *Scientific American.* June.

Witherspoon, P. A., and G. S. Bodvarsson, eds. 2001. *Geological Challenges in Radioactive Waste Isolation: Third Worldwide Review.* Berkeley: University of California Press.

4. How Fast the Rising Sea

Bindschalder, R. A., and Charles R. Bentley. 2002. On thin ice. *Scientific American* 287:98–105.

Brunner, R. D. 2001. Science and the climate change regime. *Policy Sciences* 34:1–33.

Crichton, M. 2003. Aliens cause global warming. The Cal Tech Michelin Lecture.

——. 2004. *State of Fear.* New York: HarperCollins.

Douglas, B. C., M. S. Kearney, and S. P. Leatherman. 2001. *Sea Level Rise: History and Consequences.* International Geophysics Series, vol. 75. New York: Academic Press.

Intergovernmental Panel in Climate Change (IPCC). 2001. *Climate Change 2001: Synthesis Report.*

Kolbert, Elizabeth. 2005. The climate of man I, II, and III. *New Yorker.* April 25, May 2, May 9.

Lomborg, Bjorn. 2002. The environmentalists are wrong. *New York Times.* August 26.

Perrow, C. 1999. *Normal Accidents.* Princeton, N.J.: Princeton University Press.

5. Following a Wayward Rule

Bruun, P. 1954. *Coast Erosion and the Development of Beach Profiles.* Technical Memorandum no. 44. Washington, D.C.: Beach Erosion Board, U.S. Army Corps of Engineers.

———. 1962. Sea-level rise as a cause of shore erosion. Proceedings of the American Society of Civil Engineers. *Journal of the Waterways and Harbors Division* 88:117–130.

Cooper, J. A. G., and O. H. Pilkey. 2004. Sea level rise and shoreline retreat: Time to abandon the Bruun Rule. *Global and Planetary Change* 43:157–171.

Cowell, P., and P. Kench. 2001. The morphological response of atoll islands to sea level rise. Part 2: Application of the modified Shoreline Translation Model (STM). *Journal of Coastal Research* 34:633–644.

Dean, R. G. 1990. Equilibrium beach profiles: Characteristics and applications. *Journal of Coastal Research* 7 (1): 53–84.

Leatherman, S. P. 1991. Modeling shore response to sea-level rise on sedimentary coasts. *Progress in Physical Geography* 14:447–467.

———. 2001. Social and economic costs of sea level rise. In B. C. Douglas, M. S. Kearney, and S. P. Leatherman, eds., *Sea Level Rise: History and Consequences*, 181–223. International Geophysics Series, vol. 75. New York: Academic Press.

Leatherman, S. P., K. Zhang, and B. C. Douglas. 2000. Sea level rise shown to drive coastal erosion. *EOS* 81:55–57.

Nicholls, R. J., and S. P. Leatherman, eds. 1994. Potential impacts of accelerated sea-level rise on developing countries. *Journal of Coastal Research*. Special Issue 14.

Nicholls, R. J., S. P. Leatherman, K. C. Dennis, and C. R. Volonte. 1994. Impacts and responses to sea-level rise: Qualitative and quantitative assessments. *Journal of Coastal Research*. Special Issue 14:26–43.

Pilkey, O. H., and J. A. G. Cooper. 2004. Society and sea level rise. *Science* 303:1781–1782.

Pilkey, O. H., and T. W. Davis. 1987. An analysis of coastal recession models: North Carolina coast. In D. Nummedal, O. H. Pilkey, and J. D. Howard, eds., *Sea-Level Fluctuation and Coastal Evolution*, 59–68. Tulsa, Okla.: Society of Economic Paleontologists and Mineralogists.

Riggs, S. R., W. J. Cleary, and S. W. Snyder. 1995. Influence of inherited geologic framework on barrier shoreface morphology and dynamics. *Marine Geology* 126:213–234.

Rosen, P. S. 1978. A regional test of the Bruun Rule on shoreline erosion. *Marine Geology* 26:M7–M16.

SCOR Working Group 89. 1991. The response of beaches to sea-level changes: A review of predictive models. *Journal of Coastal Research* 7 (3): 895–921.

Zhang, K., B. C. Douglas, and S. P. Leatherman. 2004. Global warming and coastal erosion. *Climatic Change* 1–18.

6. Beaches in an Expected Universe

Bodge, K. R., and N. C. Kraus. 1991. Critical examination of longshore transport rate magnitudes. *Coastal Sediments* 91:139–155.

Carter, R. W. G. 1988. *Coastal Environments*. London: Academic Press.

Cooper, J. A. G., and O. H. Pilkey. 2004. Longshore drift: Trapped in an expected universe. *Journal of Sedimentary Research* 174:599–606.

Dean, R. G., and R. A. Dalrymple. 2002. *Coastal Processes with Engineering Applications*. Cambridge: Cambridge University Press.

Haff, P. 1996. Limitations on predictive modeling in geomorphology. In B. L. Rhodes and C. E. Thorn, eds., *The Scientific Nature of Geomorphology*, 337–358. Proceedings of the 27th Binghamton Symposium in Geomorphology. New York: John Wiley.

Hanson, H., and N. C. Kraus. 1989. GENESIS: A generalized shoreline change numerical model. *Journal of Coastal Research* 5:1–27.

Komar, P. D. 1988. *Beach Processes and Sedimentation*. 2d ed. Upper Saddle River, N. J. : Prentice Hall.

Komar, P., and D. Inman. 1970. Longshore sand transport on beaches. *Journal of Geophysical Research* 75:5914–5927.

Oreskes, N., K. Shrader-Frechette, and K. Belitz. 1994. Verification, validation, and confirmation of numerical models in the earth sciences. *Science* 263:641–646.

Perrow, C. 1999. *Normal Accidents*. Princeton, N.J.: Princeton University Press.

Pilkey, O. 2000. What you know can hurt you: Predicting the behavior of nourished beaches. In D. Sarewitz, R. A. Pielke, and R. Byerly, eds., *Prediction, Science, Decision Making, and the Future of Nature*, 159–184. Washington, D.C.: Island Press.

Pilkey, O., R. S. Young, S. Riggs, A. Smith, H. Hu, and W. Pilkey. 1993. The concept of shoreface profile of equilibrium: A critical review. *Journal of Coastal Research* 9:255–278.

Seymour, R. J., and A. L. Higgins. 1978. Continuous estimation of longshore sand transport. In *Coastal Zone '78*, 3:2308–2318. New York: American Society of Civil Engineers.

Thieler, E. R., O. H. Pilkey, R. S. Young, D. M. Bush, and F. Chei. 2000. The use of mathematical models to predict beach behavior for U.S. coastal engineering. *Journal of Coastal Research* 16 (4): 48–70.

USACE. 2000. *Final Feasibility Report and Environmental Impact, Dare County Beaches (Bodie Island Portion), Dare County, North Carolina, Wilmington District, Wilmington, NC*.

———. 2001. *Final Supplement III, Environmental Impact Statement, Manteo (Shallowbag) Bay, Dare County, North Carolina, Wilmington District, Wilmington, NC.*

Wang, P., N. C. Kraus, and R. A. Davis. 1998. Total longshore sediment transport rate in the surf zone: Field measurements and empirical predictions. *Journal of Coastal Research* 14:269–282.

Wright, L. D. 1995. *Morphodynamics of Inner Continental Shelves.* Boca Raton, Fla.: CRC Press.

Young, R. S., O. H. Pilkey, D. M. Bush, and E. R. Thieler. 1997. A discussion of the Generalized Model for Simulating Shoreline Change (GENESIS). *Journal of Coastal Research* 11:875–886.

7. Giant Cups of Poison

Eggleston, J. R., and S. A. Rojstaczer. 2000. Can we predict subsurface mass transport? *Environmental Science and Technology* 34:4010–4017.

Johnson, K. 1996. Geochemical model of water quality in pit lake in oxidized tuff, McDonald Gold Mine. Unpublished consultants' report. Johnson Environmental Concepts.

Kempton, J., W. Locke, D. Atkins, and A. Nicholson. 2000. Probabilistic quantification of uncertainty in predicting mine pit lake water quality. *Mining Engineering* (October): 59–64.

Lewis, R. L. 1999. Predicting the steady state water quality of pit lake. *Mining Engineering* (October): 54–58.

Maest, A., J. Kuipers, C. Travers, and D. Atkins. 2005. Evaluation of methods and models used to protect water quality at hardrock mine sites: Sources of uncertainty and recommendations for improvement. Annual meeting proceedings. Salt Lake City: Society for Mining, Metallurgy, and Exploration.

Miller, G., W. Lyons, and A. Davis. 1996. Understanding the water quality of pit lakes. *Environmental Science and Technology* 30 (3).

Montana Mining Properties. 1990. Butte mining district history. *http://members. aol.com/MontanaMining/DistrictHistoryPage.htm.*

Moran, Robert. 2000. Is the number to your liking? Water quality predictions in mining impact studies. In D. Sarewitz, R. A. Pielke, and R. Byerly, eds., *Prediction, Science, Decision Making and the Future of Nature,* 185–198. Washington, D.C.: Island Press.

Myers, Tom. 1997. Testimony to House Subcommittee on Energy and Mineral resources. September 22.

National Academy of Sciences. 1990. *Ground Water Models: Scientific and Regulatory Applications.* Washington, D.C.: National Academies Press.

National Research Council. 1999. *Hardrock Mining on Federal Lands.* Washington, D.C.: National Academies Press.

——. 2005. *Superfund and Mining Megasites: Lessons from the Coeur d'Alene River Basin.* Washington, D.C.: National Academies Press.

Shafer and Associates. 1995. Post-closure mine lake water balance and hydrochemical assessment for the Seven-Up Joint venture. Unpublished report. Bozeman, Mont.: Shafer and Associates.

Thorne, Christopher. 2002. Berkeley Pit clean up pact reached. *Missoulian.* March 27.

Wilkinson, C. R. 1994. Hydrogeologic information and ground water modeling. In K. Baker and D. Herson, eds., *Bioremediation.* New York: McGraw-Hill.

8. Invasive Plants

Animal and Plant Health Inspection Service. 2002. *Weed risk assessment.* Version 5.2. Riverdale, Md.

Fritts, T. H., and G. H. Rodda. 1998. The role of introduced species in the degradation of island ecosystems: A case history of Guam. *Annual Review of Ecology and Systematics* 29:113–114.

Goodwin, B. J., A. J. McAllister, and L. Fahrig. 1999. Predicting invasiveness of plant species based on biological information. *Conservation Biology* 13: 422–426.

Higgins, S. I., D. M. Richardson, and R. M. Cowling. 2001. Validation of a spatial simulation model of a spreading alien plant population. *Journal of Applied Ecology* 38:571–584.

Holm, L. G., D. L. Plunknett, and J. V. Pancho. 1977. *The World's Worst Weeds: Distribution and Biology.* Melbourne, Fla.: Krieger.

MacArthur, R. H., and E. O. Wilson. 1967. *The Theory of Island Biogeography.* Princeton, N.J.: Princeton University Press.

Mack, R. N. 2003. Global plant dispersal, naturalization, and invasion: Pathways, modes, and circumstances. In G. M. Ruiz and J. T. Carlton, eds., *Invasive Species: Vectors and Management Strategies,* 3–30. Washington, D.C.: Island Press.

Mack, R. N., D. Simberloff, W. M. Lonsdale, H. Evans, M. Clout, and F. A. Bazzaz. 2000. Biotic invasions: Causes, epidemiology, global consequences, and control. *Ecological Applications* 10:689–710.

National Research Council. 2002. *Predicting Invasions of Nonindigenous Plants and Plant Pests.* Washington, D.C.: National Academies Press.

Ruiz, G. R., and L. T. Carlton, eds. 2003. *Invasive Species: Vectors and Management Strategies*. Washington, D.C.: Island Press.

Sutherst, R. W., G. F. Maywald, T. Yonow, and P. M. Stevens. 2000. *CLIMEX User Guide: Predicting the Effects of Climate on Plants and Animals*. Collingwood, Victoria, Australia: CSIRO Publishing.

Williamson, M. H. 1996. *Biological Invasions*. London: Chapman and Hall.

9. A Promise Unfulfilled

Allen, M. R., and W. J. Ingram. 2002. Constraints on future changes in climate and the hydrologic cycle. *Nature* 419:224–232.

Bobbitt, P. 2003. *The Shield of Achilles: War, Peace, and the Course of History*. New York: Knopf.

Chessire, J. H., and A. J. Surrey. 1975. World energy sources and limitations of computer modeling. *Long Range Planning* 8:60.

Chwastiak, M. 2001. Contradiction between representation and reality, planning, programming, and budgeting and the Vietnam War: A review. www.commerce.adelaide.edu.au/apira/papers/Chwastiak152.pdf.

Edwards, P. 1995. *The Closed World: Computers and the Politics of Discourse in Cold War America*. Cambridge, Mass.: MIT Press.

Haff, P. K. 1996. Limitations on predictive modeling in geomorphology. In B. L. Rhoads and C. E. Thorn, eds., *The Scientific Nature of Geomorphology*, 337–358. Proceedings of the 27th Binghamton Symposium in Geomorphology. New York: John Wiley.

Kahn, H. *On Thermonuclear War*. Westport, Conn.: Greenwood Press.

Lewis, C. S. 1952. *Mere Christianity*. Glasgow: William Collins.

Moran, Robert. 2000. Is the number to your liking? Water quality predictions in mining impact studies. In D. Sarewitz, R. A. Pielke, and R. Byerly, eds., *Prediction, Science, Decision Making, and the Future of Nature*, 185–198. Washington, D.C.: Island Press.

Pilkey, O. 2000. What you know can hurt you: Predicting the behavior of nourished beaches. In D. Sarewitz, R. A. Pielke, and R. Byerly, eds., *Prediction, Science, Decision Making, and the Future of Nature*, 159–184. Washington, D.C.: Island Press.

Sherden, W. A. 1998. *The Fortune Sellers*. New York: John Wiley.

Smil, Vaclav. 2003. *Energy at the Crossroads*. Cambridge, Mass.: MIT Press.

Tetlock, P. 2005. *Expert Political Judgment: How Good Is It? How Can We Know?* Princeton, N.J.: Princeton University Press.

index